Indigenous Rights and Water Resource Management

In an era of climate change, the need to manage our water resources effectively for future generations has become an increasingly significant challenge. Indigenous management practices have been successfully used to manage inland water systems around the world for thousands of years, and Indigenous people have begun calling for a greater role in the management of water resources. As First Peoples and as holders of important knowledge of sustainable water management practices, they regard themselves as custodians and rights holders, deserving of a meaningful role in decision-making. This book argues that a key (albeit not the only) means of ensuring appropriate participation in decision-making about water management is for such participation to be legislatively mandated. To this end, the book draws on case studies in Australia and New Zealand in order to elaborate the legislative tools necessary to ensure Indigenous participation, consultation and representation in the water management landscape.

Katie O'Bryan is based in the Faculty of Law at Monash University, Melbourne.

Part of the Indigenous Peoples and the Law series

Series editors
Dr Mark A. Harris *University of British Columbia, Canada*
Professor Denise Ferreira da Silva *University of British Columbia, Canada*
Dr Claire Charters *University of Auckland, New Zealand*
Dr Glen Coulthard *University of British Columbia, Canada*

For information about the series and details of previous and forthcoming titles, see www.routledge.com/law/series/INDPPL

Indigenous Rights and Water Resource Management

Not Just Another Stakeholder

Katie O'Bryan

Routledge
Taylor & Francis Group
a GlassHouse book

First published 2019
by Routledge
2 Park Square, Milton Park, Abingdon, Oxon OX14 4RN

and by Routledge
52 Vanderbilt Avenue, New York, NY 10017

First issued in paperback 2020

A GlassHouse book

Routledge is an imprint of the Taylor & Francis Group, an informa business

British Library Cataloguing in Publication Data
A catalogue record for this book is available from the British Library

Library of Congress Cataloging-in-Publication Data
A catalog record has been requested for this book

ISBN 13: 978-0-367-66485-5 (pbk)
ISBN 13: 978-0-8153-7542-5 (hbk)

Typeset in Galliard
by Taylor & Francis Books

Contents

Acknowledgements

This book developed out of my PhD dissertation, which I commenced after working for over a decade as a native title lawyer in both Western Australia and Victoria. Accordingly, I owe a huge debt of gratitude to my supervisors, Melissa Castan and David Yarrow, for their expert guidance, advice and knowledge, and whose ongoing friendship I value enormously. I would also like to thank Jacinta Ruru and Maureen Tehan for their excellent feedback and for encouraging me to publish my research. My heartfelt thanks also to Donna Craig for being such a wonderful mentor, and for giving me the courage to undertake my doctorate.

To my family and friends, thank you for your unwavering encouragement and support. To my many friends and colleagues in the Law Faculty at Monash University and elsewhere, thank you for sharing my journey with me. To Rexy, thank you for helping me to maintain my physical and mental health along the way.

And, finally, thank you to my Indigenous friends and colleagues, and their communities, who shared their water stories with me over the many years I spent on (and off) country working on their native title claims. Thank you for taking me into your confidence, into your homes and onto your country, for teaching me about your unique and wonderful culture, and for motivating and inspiring me to research this vitally important issue. This book is dedicated to you.

Abbreviations

AHA	*Aboriginal Heritage Act 2006* (Vic)
AHC	Australian Heritage Council
AIATSIS	Australian Institute of Aboriginal and Torres Strait Islander Studies
ALA	*Aboriginal Lands Act 1971* (Vic)
ALRC	Australian Law Reform Commission
ATSIC	Aboriginal and Torres Strait Islander Commission
BCC	Basin Community Committee
BDAC	Biological Diversity Advisory Committee
CBD	Convention on Biological Diversity
CHMP	cultural heritage management plan
CMA	Catchment Management Authority
COAG	Council of Australian Governments
CRT	Critical Race Theory
DEPI	Department of Environment and Primary Industries
DELWP	Department of Environment, Land, Water and Planning
DSE	Department of Sustainability and Environment
ECOSOC	Economic and Social Council
EES	Environmental Effects Statement
EMRIP	Expert Mechanism on the Rights of Indigenous Peoples
FPIC	free, prior and informed consent
FVTOC	Federation of Victorian Traditional Owner Corporations
GEC	Guardians Establishment Committee
GRSWS	Gippsland Region Sustainable Water Strategy
HPA	heritage protection authority
HPO	heritage protection order
IAC	Indigenous Advisory Committee
ICCPR	International Covenant on Civil and Political Rights
ICESCR	International Covenant on Economic, Social and Cultural Rights
ILO	International Labour Organization
ILUA	indigenous land use agreement
JMA	joint management agreement
JMP	joint management plan

LUA	land use activity
LUAA	land use activity agreement
LTWRA	long term water resource assessment
LWF	Land and Water Forum
MDB	Murray–Darling Basin
MDBA	Murray–Darling Basin Authority
MDBC	Murray–Darling Basin Commission
MLDRIN	Murray and Lower Darling Rivers Indigenous Nations
MMP	mixed-member proportional
MNES	Matter(s) of National Environmental Significance
NAILSMA	North Australian Indigenous Land and Sea Management Alliance
NBAN	Northern Basin Aboriginal Nations
NNTT	National Native Title Tribunal
NPS	national policy statement
NPS-FM	National Policy Statement for Freshwater Management
NRA	natural resource agreement
NRM	natural resource management
NSESD	National Strategy for Ecologically Sustainable Development
NTA	*Native Title Act 1993* (Cth)
NTSV	Native Title Services Victoria
NWC	National Water Commission
NWI	National Water Initiative
PSGE	post settlement governance entity
RAP	registered Aboriginal party
RCS	regional catchment strategy
RLSG	Rotorua Lakes Strategy Group
RMA	*Resource Management Act 1991* (NZ)
RMLR	resource management law reforms
RNTBC	registered native title body corporate
RSA	recognition and settlement agreement
RWS	regional waterway strategy
SWS	Sustainable Water Strategy
TALT	Te Arawa Lakes Trust
TOLMA	traditional owner land management agreement
TOLMB	Traditional Owner Land Management Board
TOGE	traditional owner group entity
TOSA	*Traditional Owner Settlement Act 2010* (Vic)
TSSC	Threatened Species Scientific Committee
UNDRIP	United Nations Declaration on the Rights of Indigenous People
VCAT	Victorian Civil and Administrative Tribunal
VCMC	Victorian Catchment Management Council
VEWH	Victorian Environmental Water Holder
VNTSF	Victorian Native Title Settlement Framework
VTOLJG	Victorian Traditional Owner Land Justice Group
WCO	water conservation order

WGIP	Working Group on Indigenous Populations
WMS	waterway management strategy
WRA	Waikato River Authority
WRP	water resource plan
WRRT	Waikato Raupatu River Trust
WSPA	water supply protection area

Introduction

From the dry creek beds and rock holes in the Goldfields region in Western Australia to the lakes and flowing waters of East Gippsland and the drought-stricken rivers of the Wimmera region in Victoria, my time as a native title lawyer exposed me to the many and varied ways that Indigenous Australians value water in all its forms. In almost every meeting, every field trip, every witness statement and every negotiation I was involved in, water featured in one way or another – culturally, spiritually, physically, socially, economically. It was abundantly clear that water is the lifeblood of Indigenous communities. But it was also clear that the importance for Indigenous Australians of having a meaningful say in how it should be managed was not well understood by either government or the wider community.

Indigenous management practices have, however, been successfully used to manage inland water systems in Australia (and elsewhere) for many thousands of years. In the current era of climate change, the need to manage our water resources effectively for future generations has become an increasingly significant challenge. It is therefore incumbent upon non-Indigenous people to recognise the value of Indigenous perspectives on water management if we are to meet this challenge into the future.

Indigenous people have, for many years, been calling for a greater role in the management of water resources. As First Peoples and as holders of important knowledge of sustainable water management practices, they regard themselves as custodians and *sui generis* rights holders, therefore deserving of a meaningful role in decision-making. 'We're not just another stakeholder to be consulted and ticked off a checklist' was a sentiment I heard with regular occurrence.

A key means of ensuring meaningful participation in decision-making about water management is for such participation to be legislatively mandated. Without legislatively recognised rights to participate, there is a danger that Indigenous interests will continue to be marginalised, resulting in valuable Indigenous water management practices being overlooked in the management and control of water resources and further entrenching the dispossession of Indigenous people from their land and waters.

This book considers this issue by investigating three jurisdictions: Australia, and within Australia, Victoria, and Aotearoa New Zealand. And from these case studies are drawn lessons of wider value for other common law jurisdictions in which

Indigenous people have been asserting their inherent rights to participate in the management of water resources on their traditional lands.

Overview

In Part A (Chapter 1) I explain why this field is being investigated, for which I have identified three main rationales arising out of Indigenous Peoples' status as First Peoples. First, there is value for everyone in recognising Indigenous knowledge and experience. Second, there are expectations arising from international standards such as the *United Nations Declaration on the Rights of Indigenous Peoples*, expectations frequently expressed by many Indigenous people with whom I have had contact over the years. And, finally, there is an emerging body of Indigenous legal theory which argues for engagement with Indigenous perspectives when examining laws which affect Indigenous people. I then clarify how non-Indigenous researchers and academics like me are able to assist. Finally, the importance of legal recognition is elucidated.

Having set the scene, in Part B I investigate the history of water law and regulation in Australia, to provide some background and context for the current state of Indigenous rights to participate in the management of water resources. An important element of this background investigation involves an analysis of the extent to which Indigenous people participated in the development of the regulatory regimes, as well as some of the factors affecting their ability to participate.

The federal nature of Australia's political and legal system necessitates an investigation into both the national and state dimensions of water regulation, the state under examination being Victoria. Victoria provides a useful case study of water regulation in the Australian states because it was at the forefront of modern water management in Australia, whose development is typical of most of the states. It is also my home state and where I have spent much of my time as a practising native title lawyer, so I know its climate and topography and am familiar with many of its Indigenous communities. And over the years I have witnessed droughts and floods, government responses and the absence of Indigenous voices in helping to manage these issues.

The method for my investigation involves identifying the historic and current statutory and policy framework for water management that developed out of the common law. I undertake an historical analysis of the written record (such as Hansard, newspapers, discussion papers, reports, historical material – both primary and secondary sources) to examine how the regulatory framework developed, looking in particular for evidence of any Indigenous involvement.

The next two parts of this book contain a legal analysis of the applicable legislation and case law. In Part C I analyse Commonwealth legislation, starting with the *Native Title Act 1993* (Cth) ('*NTA*') as an avenue for Indigenous people to participate in the management of water resources. The Victorian experience of the native title regime then provides some examples for understanding the limitations of this avenue. Having practised as a native title solicitor for over a decade, including in Victoria, it was clear that the *NTA* was an imperfect solution for recognising Indigenous water rights, and that there had to be a better way forward.

Other Commonwealth legislation impacting on water management is also analysed in Part C, namely the *Water Act 2007* (Cth) and the *Environment Protection and Biodiversity Conservation Act 1999* (Cth).

In Part D I turn to the Victorian regulatory framework, commencing with a chapter on the *Water Act 1989* (Vic) ('*Water Act 1989*'). After an analysis of the one section in the *Water Act 1989* that currently recognises Traditional Owner interests, I evaluate the opportunities the Act generally provides for input into water management, noting some amendments that have recently been proposed.

However, participation in water management is more than just an opportunity to provide input; at a much higher level it involves representation on decision-making bodies. Thus the main water governance entities under both the *Water Act 1989* and the *Catchment and Land Protection Act 1994* (Vic) are evaluated for their adequacy in providing for Indigenous representation.

Indigenous rights legislation is the main subject of the next chapter. I commence with an analysis of Victoria's early land rights legislation, and then move on to the more recent *Traditional Owner Settlement Act 2010* (Vic) ('*TOSA*'), Victoria's alternative to the *NTA* regime for the settlement of native title claims. This Act (the first of its kind in Australia) is understood to be a progressive solution to the limited ability of native title to deliver land justice in Victoria;[1] however, my analysis shows that its impact on water rights is insignificant.

An analysis of Victoria's *Aboriginal Heritage Act 2006* (Vic) is the subject of the next section of this chapter, as water sources are often sites of cultural heritage significance. Here I analyse the provisions relating to cultural heritage management plans, which are required for areas of cultural heritage sensitivity, a classification given to waterways and other water sources under the regulations. Finally, I touch on the *Heritage Rivers Act 1992* (Vic) and the *Planning and Environment Act 1987* (Vic).

A common thread that emerges is that when it comes to legislative recognition, Indigenous interests are still being treated less favourably than non-Indigenous interests with little regard for Indigenous Peoples' status as Australia's First Peoples; the pre-native title mentality has not yet been entirely expunged from the minds of our policy-makers and parliamentary representatives.

For my analysis, I draw heavily on the written record. The focus on written material and legislative/policy analysis is intended to provide a baseline understanding of these water management regimes to enable the subsequent identification of options which may be further explored in pursuing Indigenous aspirations for water management.

1 John Brumby, 'New Framework a Just Approach to Native Title' (Media Release, 28 July 2010); 'Vic Title Claims to Settle out of Court', *The Age* (online), 4 June 2009 <http://news.theage.com.au/breaking-news-national/vic-title-claims-to-settle-out-of-court-20090604-bwnp.html>.

Having analysed the Commonwealth and Victorian regulatory regimes, I then turn my focus in Part E to Aotearoa New Zealand. A former British colony like Australia, Aotearoa New Zealand has been grappling with this issue from a much earlier point in time, and in the last few years has seen some interesting and innovative developments. Aotearoa New Zealand is chosen as a comparative case study because of its similar legal and common law history, and colonial context, bearing in mind some very notable differences, such as the lack of a written Constitution, its political structure as a unitary state with designated Māori parliamentary seats, and recognition of Māori by way of the *Treaty of Waitangi* and in subsequent policy and laws.[2]

Part E opens with a chapter on the background and history of Aotearoa New Zealand's water management regime, highlighting various factors that have impacted on its development, not least being the existence of the *Treaty of Waitangi*.

The next chapter evaluates two avenues of Māori participatory rights in water management. Having analysed in detail the role that native title (and the *NTA*) has played in Australia in facilitating Indigenous participation in water management, I first consider the role that native title has played for Māori in Aotearoa New Zealand. The analysis shows that the native title trajectory has been quite different in Aotearoa New Zealand, with native title rights to land having been extinguished over most of the country. However native title rights in water is still an unsettled area of law.

I then turn to the *Resource Management Act 1991* (NZ) ('*RMA*'), Aotearoa New Zealand's primary statute governing the management of water resources. At the time of its enactment it was seen as revolutionary in both the extensive participation of Māori in its development, and the subsequent recognition of Māori interests in its provisions. But investigations show that although making some inroads, the *RMA* has not lived up to its initial promise in terms of Māori participatory rights.

Treaty settlements are the subject of the final chapter in Part E. In this chapter I first outline by way of background two early settlements but then focus more closely on two of the more innovative recent settlements, namely the Waikato and Whanganui River Settlements.

With an analysis of Australia, Victoria and Aotearoa New Zealand completed, Part F then draws upon those jurisdictions in identifying common principles and proposals for reform, concluding with an outline of lessons of wider value.

Scope and concepts informing the topic of this book

This book evaluates the extent to which the legislative and regulatory regimes under analysis provide for Indigenous participation in decision-making processes relating to water management.

2 For a detailed discussion on the value of comparative analysis in the context of native title, see Simon Young, *The Trouble with Tradition: Native Title and Cultural Change* (Federation Press, 2008) ch 3.

First, a note on terminology. In this book, the term 'Indigenous' is generally used to refer to both Aboriginal and Torres Strait Islander Australians collectively, and Indigenous people more generally. When referring to Victoria's Indigenous people, I use the local self-identification term where appropriate. The term 'Traditional Owner' is also used in the Victorian context, particularly in light of the *TOSA*. The term 'Aboriginal people' (or 'Aboriginal Victorians') is also used as an alternative to Traditional Owner, as it is a term which appears in much of the legislation under analysis.[3]

There are three additional terms which require brief mention, as they inform the scope of the topic; namely participation, management and water.

Participation comes in many forms, ranging from mere notification to control,[4] but the focus here is on participation by way of consultation, and by way of representation on advisory committees and decision-making bodies. Each of these forms of participation can be seen as a rung on a ladder, with consultation being a rung at the lower end and representation on decision-making bodies at the upper end.[5]

Management is defined as 'the act or manner of managing; handling, direction, or control'.[6] In the context of this book, I do not consider questions of property rights (ownership), access to or the allocation and use of water, except to the extent that management involves making decisions about such matters. Ownership, access and allocation issues are necessarily much broader and invite questions of corporate rights and commercial uses; these are beyond the scope of this book.

Finally, this book is also confined to legislation relating to inland waters, the sea being subject to a different legislative and regulatory regime. Generally speaking, I have only delved into the distinction between groundwater and surface water where required by the context, as this is a common law and statutory construct which does not reflect Indigenous conceptions of country. Notably, this book itself can be said to perpetuate an Anglo-centric conception of water as separate from land, and indeed adopts another dichotomy. It is acknowledged that this is not consistent with Indigenous world views; however, my investigation examines Australian (and Aotearoa New Zealand) legal and regulatory regimes which themselves separate water from land, so this necessarily requires that I take this approach.

3 See, eg, *Aboriginal Heritage Act 2006* (Vic); *Constitution of Victoria Act 1975* (Vic); *Aboriginal Lands Act 1970* (Vic).

4 See, eg, Local Government New Zealand, *Co-Management: Case Studies Involving Local Authorities and Māori* (January 2007) 7. See also Sherry R Arnstein, 'A Ladder of Citizen Participation' (1969) 35(4) *Journal of the American Institute of Planners* 216. This ladder of citizen participation has been used in the context of co-management between the state and Indigenous groups; see, eg, Fikret Berkes, Peter George and Richard J Preston, 'Co-Management: The Evolution in Theory and Practice of the Joint Administration of Living Resources' (1991) 18(2) *Alternatives* 12, 12.

5 With thanks to Arnstein, above n 4, for the ladder analogy.

6 *Macquarie Dictionary Online* (Macquarie Dictionary Publishers, 2018) <www.macquariedictionary.com.au>.

Part A

The international context

1 Exploring Indigenous participation – why is it important?

Holders of valuable knowledge

It is uncontroversial that there is a wealth of valuable knowledge held by Indigenous people about natural resource management. As noted in the 1987 report of the World Commission on Environment and Development, *Our Common Future*:

> [Indigenous] communities are the repositories of vast accumulations of traditional knowledge and experience that links humanity with its ancient origins. Their disappearance is a loss for the larger society, which could learn a great deal from their traditional skills in sustainably managing very complex ecological systems.[1]

More specific to water management, and in light of the case studies in this book, Australia's National Water Commission acknowledged in 2012 that:

> Indigenous Australians have managed their lands and waters sustainably for thousands of generations. Through their spiritual, cultural and customary connections to the landscape, they have acquired a deep knowledge and understanding of Australia's water systems.[2]

Despite the devastating effects of colonisation, many Indigenous Australians still hold valuable knowledge of ecosystem management,[3] as do Indigenous people in other colonised countries. It is therefore vitally important that Indigenous people are able to maintain and use their knowledge by participating in water management decision-making processes.

1 *Report of the World Commission on Environment and Development*, UN GA, 42nd sess, UN Doc A/42/427 (4 August 1987) annex 119 para 74.
2 National Water Commission, *Position Statement: Indigenous Access to Water Resources* (June 2012) 1.
3 See, eg, *Gippsland Region Sustainable Water Strategy* (Department of Sustainability and Environment, November 2011) 81.

Expectations arising from international law

The construction and recognition of Indigenous rights in international law is plagued by many issues both in their development and implementation. These include the related concepts of state sovereignty and territorial integrity,[4] the collective nature of Indigenous rights,[5] definitions of Indigenous peoples[6] and the broad meaning of self-determination,[7] to name but a few. It is beyond the scope of this book to canvass these issues, but their existence indicates the complexities involved when considering Indigenous rights.

Yet, despite these complexities, in September 2007 the United Nations General Assembly adopted the *UN Declaration on the Rights of Indigenous Peoples* (*UNDRIP*),[8] the pre-eminent international document on Indigenous rights. The involvement of Indigenous people in its development was crucial to its success.

The development of the *UNDRIP* commenced in 1982 with the establishment of the Working Group on Indigenous Populations ('WGIP').[9] The WGIP was a major development in Indigenous rights at the international level. Arising out of the recommendations of the seminal Martínez Cobo Study,[10] its mandate (as

4 See, eg, Austin L Parrish, 'Changing Territoriality, Fading Sovereignty, and the Development of Indigenous Rights' (2007) 31 *American Indian Law Review* 291. See also Megan Davis, 'Indigenous Struggles in Standard-setting: The United Nations Declaration on the Rights of Indigenous Peoples' (2008) 9(2) *Melbourne Journal of International Law* 439, 457–61, in the context of the drafting of the *UNDRIP*.

5 See, eg, Will Kymlicka, *Multicultural Citizenship: A Liberal Theory of Minority Rights* (Oxford University Press, 1995) 34–48; Leonard Hammer, 'Indigenous Peoples as a Catalyst for Applying the Human Right to Water' (2003) 10 *International Journal on Minority and Group Rights* 131; Odette Mazel, 'The Evolution of Rights: Indigenous Peoples and International Law' (2009) 13(1) *Australian Indigenous Law Review* 140.

6 See, eg, Robert K Hitchcock, 'International Human Rights, the Environment, and Indigenous Peoples' (1994) 5 *Colorado Journal of International Environmental Law and Policy* 1; Benedict Kingsbury, 'Indigenous Peoples in International Law: A Constructivist Approach to the Asian Controversy' (1998) 92 *American Journal of International Law* 414; Mazel, above n 5, 147–8; Patrick Thornberry, *Indigenous Peoples and Human Rights* (Manchester University Press, 2005) 33–60.

7 See, eg, Raidza Torres, 'The Rights of Indigenous Populations: The Emerging International Norm' (1991) 16 *Yale Journal of International Law* 127; Hitchcock, above n 6; Benedict Kingsbury, 'Reconciling Five Competing Conceptual Structures of Indigenous Peoples' Claims in International and Comparative Law' (2001) 34 *New York University Journal of International Law and Politics* 189; Davis, above n 4; Mazel, above n 5; Melissa Castan, 'DRIP Feed: The Slow Reconstruction of Self-Determination for Indigenous Peoples' in Sarah Joseph and Adam McBeth (eds), *Research Handbook on International Human Rights Law* (Edward Elgar, 2010) 492.

8 GA Res 61/295, UN GAOR, 61st sess, 107th plen mtg, Supp No 49, UN Doc A/RES/61/295 (13 September 2007).

9 ECOSOC Res 1982/34, UN ESCOR, 1st sess, 28th plen mtg, Supp No 1, UN Doc E/1982/82 (7 May 1982).

10 José R Martínez Cobo, *Study of the Problem of Discrimination Against Indigenous Populations*, UN Docs E/CN.4/Sub.2/1986/7 and E/CN4/Sub.2/1986/7/Add.4 (1986).

refined in 1985) included the drafting of a declaration on the rights of Indigenous peoples.[11] In fulfilling its mandate, the WGIP was required to review information and materials provided by various government and non-government groups, but 'particularly those of indigenous peoples'.[12]

Although not one of the inaugural members of WGIP, Australia participated in the first session as an observer, along with Aotearoa New Zealand, Canada and the United States.[13] In the report of that session, it was noted that '[m]ost representatives of NGOs and of Indigenous Populations as well as some government observers stressed the importance of consultation in formulating and implementing national and international standards [relating to the recognition of Indigenous rights]'.[14] The report also noted that

> several NGOs and Indigenous representatives emphasized the need for consultation with the indigenous populations before making decisions that might affect the rights to their lands, to their natural resources and to develop their natural environment within the framework of their traditional way of life, as well as any decision concerning their status or other matters of their concern.[15]

These statements have been highlighted to show the emphasis being placed on consultation with Indigenous people.

A draft declaration on the rights of Indigenous peoples was produced by WGIP. But the task of developing the final declaration was eventually undertaken not by WGIP but by a working group of the Commission on Human Rights.[16] The working group put special processes in place to ensure that Indigenous people were able to participate in the declaration's development.[17]

On 13 September 2007 the *UNDRIP* was finally adopted by the General Assembly. According to Megan Davis, who participated in its drafting, it was 'a triumph for indigenous peoples after persevering for more than 20 years to secure an international instrument aimed at recognising the distinct cultural rights of indigenous peoples and providing redress for the injustice of dispossession'.[18] The *UNDRIP* was adopted by a majority of 143 states, with four votes against, and 11 abstentions. Notably, the four countries who voted against the *UNDRIP* were

11 HRC Sub-Commission Res 1985/22, 38th sess, 37th mtg, UN Doc E/CN.4/1985/5; E/CN.4/Sub.2/1985/57 (4 November 1985).

12 ECOSOC Res 1982/34, UN ESCOR, 1st sess, 28th plen mtg, Supp No 1, UN Doc E/1982/82 (7 May 1982).

13 WGIP, *Report of the Working Group on Indigenous Populations on its First Session*, UN Doc E/CN.4/Sub.2/1982/33 (25 August 1982) [4].

14 Ibid [73].

15 Ibid [84].

16 Castan, above n 7, 497; Davis, above n 4, 446.

17 Heather McRae et al, *Indigenous Legal Issues, Commentary and Materials* (Lawbook, 4th edn, 2009) 662; Davis, above n 4, 446–7.

18 Davis, above n 4, 440.

Australia, Aotearoa New Zealand, Canada and the United States, all of which have now endorsed the *UNDRIP*. [19]

Much has been written on the evolution of Indigenous rights; through general human rights instruments, Indigenous specific instruments such as *UNDRIP* as well as via international environmental law and customary international law.[20] Rather than repeat what is already well documented, the following section will focus on participation, consultation and the overarching principle of free, prior and informed consent.

Free, prior and informed consent, consultation and participation in decision-making

Various international instruments acknowledge that Indigenous people have a right to participate in decision-making in matters that affect their rights, the most recent and most significant being the *UNDRIP*. [21] These are outlined below.

19 Jenny Macklin, 'Statement on the United Nations Declaration of Indigenous Peoples, Parliament House, Canberra, 3 April 2009' <www.un.org/esa/socdev/unpfii/documents/Australia_official_statement_endorsement_UNDRIP.pdf>; New Zealand, *Parliamentary Debates*, 20 April 2010, 10229 (Simon Power, Minister of Justice); Aboriginal Affairs and Northern Development Canada, 'Canada's Statement of Support on the United Nations Declaration on the Rights of Indigenous Peoples, November 12, 2010' (30 July 2012) <www.aadnc-aandc.gc.ca/eng/1309374239861/1309374546142>; US Department of State, *Announcement of US Support for the United Nations Declaration on the Rights of Indigenous Peoples* (12 January 2100) <https://2009-2017.state.gov/s/srgia/154553.htm>.

20 See, eg, S James Anaya, *International Human Rights and Indigenous Peoples* (Aspen, 2009); S James Anaya, *Indigenous Peoples in International Law* (Oxford University Press, 2nd edn, 2004); Cherie Metcalf, 'Indigenous Rights and the Environment: Evolving International Law' (2003) 35 *Ottawa Law Review* 101; S J Anaya, 'The Emergence of Customary International Law Concerning the Rights of Indigenous Peoples' (2005) 12 *Law and Anthropology* 127; John D Smelcer, 'Using International Law More Effectively to Secure and Advance Indigenous Peoples' Rights: Towards Enforcement in U.S. and Australian Domestic Courts' (2006) 15 *Pacific Rim Law and Policy Journal* 301; David H Getches, 'Indigenous Peoples' Rights to Water Under International Norms' (2005) 16 *Colorado Journal of International Environmental Law and Policy* 259; Mazel, above n 5; Patrick Macklem, 'Indigenous Recognition in International Law: Theoretical Observations' (2008) 30 *Michigan Journal of International Law* 177; Thornberry, above n 6.

21 GA Res 61/295, UN GAOR, 61st sess, 107th plen mtg, Supp No 49, UN Doc A/RES/61/295 (13 September 2007) art 18. See also *Convention on Biological Diversity*, opened for signature 5 June 1992, 1760 UNTS 79 (entered into force 29 December 1993) art 8(j); ILO Convention No 169, *Concerning Indigenous and Tribal Peoples in Independent Countries*, opened for signature 27 June 1989 (entered into force 5 September 1991); *Report of the United Nations Conference on Environment and Development*, UN Doc A/CONF.151/26 (Vol 1) (3–14 June 1992) annex 1 cl 22 ('*Rio Declaration*'); *Report of the United Nations Conference on Environment and Development*, UN Doc A/CONF.151/26 (Vol 1) (3–14 June 1992) annex II Ch 26, cl 18.9.c ('*Agenda 21*'); Human Rights Committee, 50th sess, *General Comment No 23: The Rights of Minorities* (Article 27), UN Doc CCPR/C/21/Rev.1/Add.5 (8 April 1994)

Participation, in this context, ranges from representation on decision-making structures to consultation of varying levels, down to mere notification. The overarching principle governing the adequacy of each form is that of free, prior and informed consent ('FPIC'), described by the Expert Mechanism on the Rights of Indigenous Peoples ('EMRIP') as follows:

> The element of 'free' implies no coercion, intimidation or manipulation; 'prior' implies that consent is obtained in advance of the activity associated with the decision being made, and includes the time necessary to allow indigenous peoples to undertake their own decision-making processes; 'informed' implies that indigenous peoples have been provided all information relating to the activity and that that information is objective, accurate and presented in a manner and form understandable to indigenous peoples; 'consent' implies that indigenous peoples have agreed to the activity that is the subject of the relevant decision, which may also be subject to conditions.[22]

EMRIP then proceeds to describe the relationship between FPIC, participation and consultation:

> The duty of the State to obtain indigenous peoples' free, prior and informed consent entitles indigenous peoples to effectively determine the outcome of decision-making that affects them, not merely a right to be involved in such processes. Consent is a significant element of the decision-making process obtained through genuine consultation and participation. Hence, the duty to obtain the free, prior and informed consent of indigenous peoples is not only a procedural process but a substantive mechanism to ensure the respect of indigenous peoples' rights.[23]

Thus, appropriate consultation and participation in decision-making will lead to the FPIC of Indigenous people for activities being undertaken on Indigenous land and waters.

Indicators of good practice decision-making

EMRIP suggests that the most significant indicator of good practice decision-making is 'the extent to which indigenous peoples were involved in the design of the practice and their agreement to it'.[24] Other indicators of good practice decision-making identified by EMRIP include:

cl 7; Committee on the Elimination of Racial Discrimination, General Recommendation No 23: Indigenous Peoples, 51st sess, UN Doc A/52/18, annex V at 122 (18 August 1997) cl 4.

22 EMRIP, *Final Report of the Study on Indigenous Peoples and the Right to Participate in Decision-Making*, UN Doc A/HRC/18/42 (17 August 2011) 27.

23 Ibid 26.

24 Ibid 4.

the extent to which the practice:

(a) Allows and enhances indigenous peoples' participation in decision-making;
(b) Allows indigenous peoples to influence the outcome of decisions that affect them;
(c) Realizes indigenous peoples' right to self-determination;
(d) Includes, as appropriate, robust consultation procedures and/or processes to seek indigenous peoples' free, prior and informed consent.[25]

There are two aspects of these good practice decision-making indicators that I draw upon in my analysis of legislation relating to Indigenous participation in water management:

1 The involvement of Indigenous people in the *development* of the legislation containing the decision-making process; and
2 The process that *resulted* from the involvement of Indigenous people in its development and design.

Before I get to that analysis, I will briefly outline the development of Indigenous participatory rights, the duty to consult and the associated FPIC principle.

The development of the principle of free, prior and informed consent and the duty to consult

FPIC is evolving as a relatively recent, albeit fundamental, principle, but its origins can be traced back to the beginning of the modern human rights era. FPIC, the duty to consult and the right to participate in decision-making find their genesis in the right to self-determination, a right which first found expression in 1945 in the UN *Charter*, [26] and then in 1966 in the *International Covenant on Civil and Political Rights* ('*ICCPR*')[27] and the *International Covenant on Economic, Social and Cultural Rights* ('*ICESCR*').[28] Article 1 in both the *ICCPR* and *ICESCR* are identical and state:

> All peoples have the right of self-determination. By virtue of that right they freely determine their political status and freely pursue their economic, social and cultural development.

This terminology has been replicated in article 1 of the *UNDRIP*, but with specific reference to Indigenous peoples.

25 Ibid.
26 *Charter of the United Nations* art 1(2).
27 *ICCPR*, opened for signature 16 December 1966, 999 UNTS 171 (entered into force 23 March 1976).
28 *ICESCR*, opened for signature 16 December 1966, 993 UNTS 3 (entered into force 3 November 1976).

As EMRIP notes, 'the normative international human rights framework for the collective right to participation is the right to self-determination'.[29] The Special Rapporteur has stated that the related duty to consult 'derives from the overarching right of indigenous peoples to self-determination'.[30] He goes on to explain that '[t]he right of self-determination is a foundational right, without which indigenous peoples' human rights, both collective and individual, cannot be fully enjoyed'.[31]

The meaning of self-determination is the subject of vigorous and ongoing debate which is beyond the scope of this book.[32] But at the most basic level understood and agreed to by the international community, self-determination is that which is not intended to 'dismember or impair, totally or in part, the territorial integrity or political unity of sovereign and independent States'. This is reflected in article 46(1) of the *UNDRIP*.

Although none of these early instruments refer specifically to Indigenous peoples, they have been used and interpreted over the years to confirm and protect Indigenous rights.[33] For example, article 27 of the *ICCPR* relating to the rights of minorities to enjoy their own culture has been used (with varying success) by Indigenous peoples. It was successfully invoked in the 2009 decision of the Human Rights Committee in *Poma Poma v Peru*, [34] a case which also commented on the FPIC principle[35] and, coincidentally, involved water rights.

It has been further suggested by at least one eminent commentator that the duty to consult with Indigenous peoples is now a part of customary international law as well.[36]

ILO Convention No 169, Concerning Indigenous and Tribal Peoples in Independent Countries *(1989)*

The first international instrument relating specifically to Indigenous rights was ILO Convention No 107 of 1957, *Concerning the Protection and Integration of*

29 EMRIP, *Progress Report on the Study on Indigenous Peoples and the Right to Participate in Decision-Making* (2010) UN Doc A/HRC/15/35 (23 August 2010) 9.

30 James Anaya, *Promotion and Protection of All Human Rights, Civil, Political, Economic, Social and Cultural Rights, Including the Right to Development: Report of the Special Rapporteur on the Situation of Human Rights and Fundamental Freedoms of Indigenous People,* UN Doc A/HRC/12/34 (15 July 2009) 14.

31 Ibid.

32 See above n 7.

33 For a summary of international recognition of Indigenous peoples' rights prior to World War II, see Macklem, above n 20, 187–90.

34 Human Rights Committee, *Views: Communication No 1457/2006*, 95th sess, UN Doc CCPR/C/95/D/1457/2006 (24 April 2009).

35 Ibid [7.6].

36 S James Anaya, 'Indigenous Peoples' Participatory Rights in Relation to Decisions About Natural Resource Extraction: The More Fundamental Issue of What Rights Indigenous Peoples Have in Lands and Resources' (2005) 22(1) *Arizona Journal of International and Comparative Law* 7, 7.

Indigenous and Other Tribal and Semi-Tribal Populations in Independent Countries. [37] Largely because of its assimilationist/integrationist focus, it was eventually replaced in 1989 by ILO Convention No 169, *Concerning Indigenous and Tribal Peoples in Independent Countries* ('*ILO Convention 169*').[38]

Australia was a founding member of the International Labour Organization ('ILO'), but has not ratified *ILO Convention 169.* It did, however, vote in favour of its adoption.[39] Also voting in favour of the Convention, and which have not subsequently ratified it, were Canada, the United States and Aotearoa New Zealand.[40]

The duty to consult and the right to participate in decision-making together form the cornerstone of *ILO Convention 169.* [41] Article 6 requires that in applying the provisions of the Convention, governments are required to consult with Indigenous peoples, and to establish mechanisms for their participation in decision-making, with the objective of achieving their agreement or consent.[42]

ILO Convention 169 was criticised for, among other things, the limited Indigenous involvement in its drafting,[43] and to date has had only 22 ratifications.[44] It remains significant, however, as the only international convention in force specifically relating to Indigenous rights, and is arguably supportive of the emergence of a norm of customary international law relating to Indigenous rights.[45]

The Earth Summit documents

The 1992 Earth Summit, held in Rio de Janeiro, produced a number of documents that explicitly refer to Indigenous participatory rights relating to the

37 *ILO Convention 107*, opened for signature 26 June 1957 (entered into force 2 June 1959).

38 *ILO Convention 169*, opened for signature 27 June 1989 (entered into force 5 September 1991).

39 International Labour Conference, *Provisional Record 32, Seventy-sixth Session, Geneva 1989, Thirty-sixth Sitting, Tuesday, 27 June*, 32/17.

40 Ibid 32/17, 32/18.

41 ILO, *Indigenous and Tribal Peoples' Rights in Practice: A Guide to ILO Convention No 169* (2009) 59; ILO, *Understanding the Indigenous and Tribal Peoples Convention, 1989 (No. 169) Handbook for ILO Tripartite Constituents* (2013) 11.

42 See also art 7(1).

43 Thornberry, above n 6, 339–40. See also Dalee Sambo, 'Indigenous Peoples and International Standard Setting Processes: Are State Governments Listening?' (1993) 3(1) *Transnational Law and Contemporary Problems* 13, 20–1; Howard R Berman, 'The International Labour Organization and Indigenous Peoples: Revision of *ILO Convention No 107* at the 75th Session of the International Labour Conference, 1988' (1988) 41 *The Review (International Commission of Jurists)* 48, 57 n 9; James Anaya, *Indigenous Peoples in International Law* (Oxford University Press, 2004) 59.

44 ILO, *Ratifications of C169 – Indigenous and Tribal Peoples Convention, 1989 (No. 169)* <www.ilo.org/dyn/normlex/en/f?p=NORMLEXPUB:11300:0::NO:11300:P11300_INSTRUMENT_ID:312314:NO>.

45 James Anaya, 'The Emergence of Customary International Law Concerning the Rights of Indigenous Peoples' (2005) 12 *Law and Anthropology* 127, 130–1.

environment. Of particular note are the *Rio Declaration*, [46]*Agenda 21* [47] and the *Convention on Biological Diversity* ('*CBD*').[48]

The importance of Indigenous participation is affirmed in clause 22 of the *Rio Declaration*, which states: 'Indigenous People ... have a vital role in environmental management ... [accordingly] ... States should ... enable their effective participation in the achievement of sustainable development.'

Agenda 21 devotes an entire chapter (ch 26) to 'Recognising and Strengthening the Role of Indigenous People and their Communities'. This chapter emphasises partnership, active participation and consultation with Indigenous peoples.[49] Chapter 18 is also important, as it relates to water and water resources. Notably, para 18.9.c. calls for the participation of Indigenous people in the management of water resources.

The *CBD* also acknowledges participatory rights of Indigenous people in relation to the use and application of traditional ecological knowledge, namely in art 8 (j). Official UN records indicate that Australia took an active role in the drafting of the *CBD*, [50] including in relation to what was eventually to become article 8(j).[51] But there is no evidence in the UN records to suggest that Indigenous people participated in the development of the *CBD* or, more specifically, the development of article 8(j). Nonetheless, it would appear that there was some limited consultation.[52] This limited Indigenous consultation is reflected in the minimal recognition of Indigenous interests which were eventually included in the *CBD*. The *CBD* does, however, contain one of the few early references to prior informed consent, although not specifically in relation to Indigenous people.[53]

46 *Report of the United Nations Conference on Environment and Development*, UN Doc A/CONF.151/26 (Vol 1) (3–14 June 1992) annex 1.
47 Ibid annex II.
48 *CBD*, opened for signature 5 June 1992, 1760 UNTS 79 (entered into force 29 December 1993).
49 See, eg, para 26.5.a.
50 Intergovernmental Negotiating Committee for a Convention on Biological Diversity ('INCCBD'), *Report of the Intergovernmental Negotiating Committee for a Convention on Biological Diversity on the Work of its Fourth Negotiating Session/Second Session*, UN Doc UNEP/Bio.Div/N4-INC.2/5 (2 October 1991) 5 [27], 12 [66], 15 [95]; INCCBD, *Report of the Intergovernmental Negotiating Committee for a Convention on Biological Diversity on the Work of its Third Session/Fifth Negotiating Session*, UN Doc UNEP/Bio.Div/N5-Inc.3/4 (4 December 1991) 2 [8], 4 [20], 8 [35], 14 [78]. INCCBD, *Report of the Intergovernmental Negotiating Committee for a Convention on Biological Diversity on the Work of its Sixth Negotiation Session/Fourth Session of INC*, UN Doc UNEP/Bio.Div/N6-INC.4/4 (18 February 1992) 3 [8], 8 [39], 13 [40].
51 INCCBD, *Report of the Intergovernmental Negotiating Committee for a Convention on Biological Diversity on the Work of its Fourth Negotiating Session/Second Session*, UN Doc UNEP/Bio.Div/N4-INC.2/5 (2 October 1991) 22, n 7.
52 International Alliance of the Indigenous Peoples of the Tropical Forests, 'The Biodiversity Convention: The Concerns of Indigenous Peoples' (1996) 1(4) *Australian Indigenous Law Reporter* 731, 731.
53 *CBD* art 15(5). Prior to the *CBD*, the only treaty reference to prior informed consent was contained in the *Convention on the Control of Transboundary Movements of*

The *CBD* (and art 8(j)) is of particular note because it is one of the few relevant treaties that has been ratified by Australia, and has been implemented domestically (in part) in the *Environment Protection and Biodiversity Conservation Act 1999* (Cth) ('*EPBC Act*').[54]

UN Declaration on the Rights of Indigenous Peoples

The development of the *UNDRIP* was outlined earlier. The duty to consult and the FPIC principle are specifically acknowledged in the *UNDRIP*. [55] It also contains numerous articles which have been cited by Indigenous people around the world, including Indigenous Australians, in support of their aspirations for water,[56] such as articles 18, 19, 25, 26 and 32.

The *UNDRIP*, although not binding on signatories, is undoubtedly the most comprehensive of all the international documents relating to Indigenous rights, and carries significant moral authority.

All of the above rights may be used to support calls for better participation by Indigenous people in water management, but they are not enforceable in most common law countries, such as Australia, unless they have been incorporated into domestic legislation.[57] Enforcement mechanisms and their associated remedies in international law generally have limited coercive power, relying largely on what is commonly termed 'naming and shaming',[58] which may have little effect on, or elicit a belligerent response from, the government of the country in question.[59]

Hazardous Wastes and Their Disposal, opened for signature 20 March 1989, 1763 UNTS 57 (entered into force 5 May 1992): Lyle Glowka et al, *A Guide to the Convention on Biological Diversity* (IUCN Environmental Law Centre, 1994) 80.

54 Article 8(j) is referred to in s 305(6) of the *EPBC Act*, along with art 10(c) and art 18(4).

55 See, eg, *UNDRIP* arts 19, 32.

56 Federation of Victorian Traditional Owner Corporations, *Victorian Traditional Owner Water Policy Statement 2014* (November 2014) 6 <www.fvtoc.com.au> ('*FVTOC Water Policy Statement 2014*'). North Australian Indigenous Experts Water Futures Forum, *Mary River Statement* (6 August 2009); Northern Australia Indigenous Land and Sea Management Alliance, *A Policy Statement on North Australian Indigenous Water Rights* (November 2009); First Peoples' Water Engagement Council, *Policy Framework* (March 2012).

57 The situation in the US is a little more complicated; see James Crawford, *Brownlie's Principles of Public International Law* (Oxford University Press, 8th edn, 2012) 77–80.

58 See, eg, Emilie M Hafner-Burton, 'Sticks and Stones: Naming and Shaming the Human Rights Enforcement Problem' (2008) 62 *International Organization* 689; James H Lebovic and Eric Voeten, 'The Politics of Shame: The Condemnation of Country Human Rights Practices in the UNHCR' (2006) 50 *International Studies Quarterly* 861. Other enforcement mechanisms are sanctions, and military force. See, eg, Sarah Joseph and Joanna Kyriakakis, 'The United Nations and Human Rights' in S Joseph and A McBeth (eds), *Research Handbook on International Human Rights Law* (Edward Elgar, 2010) 26–34.

59 See, eg, Australia's response to the Committee on the Elimination of Racial Discrimination decision on the 1998 amendments to the *Native Title Act 1993* (Cth)

Nonetheless, international law, even if it has not been incorporated into domestic legislation, still has a role to play, although limited, in domestic legal systems. Australian courts, for example, have shown that they are not averse to using international law in relevant circumstances, the Australian High Court noting in several major cases the role that international law plays in the domestic sphere.[60] It is also an important advocacy tool for Indigenous peoples.[61]

Expectations arising from Indigenous legal theory

Despite the growing recognition of Indigenous rights in domestic and international law, there has been limited attention paid to a theoretical basis for those rights. Indigenous legal theorists are rightly calling for better engagement with Indigenous perspectives in the examination, development and implementation of laws which affect Indigenous people.

Indigenous legal theory examines Indigenous legal rights from an explicitly Indigenous perspective, as opposed to simply adopting (or adapting) mainstream, or even race-conscious theories such as Critical Race Theory ('CRT'), to explain an Indigenous position. An Indigenous legal theory is one 'that may borrow from other sources, but which in some central way must be connected to Indigenous communities'.[62] Canadian academic Gordon Christie points out that there can be no single Indigenous legal theory, just as there is no single non-Indigenous legal theory,[63] asking: 'Why would we assume, for example that [Indigenous peoples] would speak with a single unified voice on theoretical matters, while non-Indigenous [Western] legal theorists produce many and varied theories about the law?'[64] There are no doubt commonalities between Indigenous peoples, but each community will have its own distinct legal system from which a theory, or theories, can be derived.

There are numerous reasons why there is a need for the development of Indigenous legal theories.[65] One important rationale is that Indigenous legal theories embrace Indigenous world views which are a necessary juxtaposition to Western

(*Report of the Committee on the Elimination of Racial Discrimination*, GAOR 54th sess, 1331st mtg, Supp 18, UN Doc A/54/18, 5–7): McRae et al, above n 17, 674. See also Devika Hovell, 'The Sovereignty Stratagem: Australia's Response to UN Human Rights Bodies' (2003) 28 *Alternative Law Journal* 197.

60 See, eg, *Polites v Commonwealth* (1945) 70 CLR 60, 68–69 (Latham CJ), 77 (Dixon J), 80–81 (Williams J); *Mabo* (1992) 175 CLR 1, 42 (Brennan J); *Minister for Immigration and Ethnic Affairs v Teoh* (1995) 183 CLR 273, 287 (Mason CJ, Deane J); *Plaintiff S157/2002 v Commonwealth* (2003) 211 CLR 476, 492 (Gleeson CJ).

61 See, eg, McRae et al, above n 17, 675.

62 Gordon Christie, 'Indigenous Legal Theory: Some Initial Considerations' in Benjamin J Richardson, Shin Iman and Kent McNeil (eds), *Indigenous Peoples and the Law: A Comparative and Critical Perspective* (Hart, 2009) 195, 197.

63 Ibid 203.

64 Ibid 203.

65 Ibid 203–10; 218–9.

(Eurocentric) legal theories that underpin the dominant Western legal system, the very instruments of Indigenous dispossession and oppression.[66] Unless Indigenous scholars enter the theoretical space, and we embrace their role in that discourse, these Western legal theories will continue to be agents of dispossession.

Benedict Kingsbury has suggested that Indigenous rights claims have been made under five different but often overlapping (and often competing) conceptual structures.[67] They include minority rights, protected by article 27 of the *ICCPR*, which Will Kymlicka has employed as a basis on which to theorise about Indigenous rights.[68] Others have turned, for example, to using CRT to support claims for Indigenous rights.[69] But very little has been written which can be described as 'Indigenous legal theory'. As Christie notes of Canada, 'there is as yet no distinct and vibrant body of scholarship identifiable as Indigenous Legal Theory'.[70] The same is just as true of Australia, and arguably of Aotearoa New Zealand as well.

Nonetheless, Canada does have a growing body of literature emerging from Indigenous legal scholars that touches on this concept of Indigenous legal theory. The earliest of the Indigenous Canadian legal scholars in this field include Mary Ellen Turpel-Lafond, Patricia Monture-Angus and Darlene Johnston, all of whom were writing in the late 1980s to mid-1990s.[71] Joining them in their endeavours are Indigenous scholars such as John Borrows, Gordon Christie, Dale Turner,

66 See, eg, Robert A Williams, Jr, *The American Indian in Western Legal Thought: The Discourses of Conquest* (Oxford University Press, 1990).

67 Kingsbury, 'Reconciling Five Competing Conceptual Structures', above n 7. Those structures are, at 190: '(1) human rights and non-discrimination claims; (2) minority claims; (3) self-determination claims; (4) historic sovereignty claims; and (5) claims as Indigenous peoples, including claims based on treaty or other agreements between Indigenous peoples and states.'

68 Kymlicka, above n 5, in particular ch 3.

69 See, eg, Robert A Williams, Jr, 'Vampires Anonymous and Critical Race Practice' (1997) 95 *Michigan Law Review* 741.

70 Christie, above n 62, 195.

71 See, eg, Aki-Kwe/Mary Ellen Turpel, 'Aboriginal Peoples and the Canadian Charter of Rights and Freedoms: Contradictions and Challenges' (1989) 10 *Canadian Woman Studies* 149; Mary Ellen Turpel, 'Indigenous Peoples' Rights of Political Participation and Self-Determination: Recent International Legal Developments and the Continuing Struggle for Recognition' (1992) 25 *Cornell International Law Journal* 579; Darlene Johnston, 'Native Rights as Collective Rights: A Question of Group Self-Preservation' (1989) 2(1) *Canadian Journal of Law and Jurisprudence 19*; Darlene M Johnston, 'A Theory of Crown Trust Towards Aboriginal Peoples' (1986) 18 *Ottawa Law Review* 307; Darlene Johnston, *The Taking of Indian Lands in Canada: Consent or Coercion?* (University of Saskatchewan Native Law Centre, 1989); Patricia A Monture, 'Ka-Nin-Geh-Heh-Gah-E-Sa-Nonh-Yah-Gah' (1988) 2 *Canadian Journal of Women and Law* 159; Patricia A Monture, 'Now that the Door is Open: First Nations and the Law School Experience' (1990) 15 *Queen's Law Journal* 179; Patricia Monture, 'Reflecting on Flint Woman' in Richard F Devlin (ed), *Canadian Perspectives on Legal Theory* (Emond Montgomery, 1991); Patricia Monture-Angus, *Thunder In My Soul: A Mohawk Woman Speaks* (Fernwood, 1995).

James (Sákéj) Youngblood Henderson and Val Napoleon. Non-Indigenous legal scholars have also played their part.[72] But it is the Indigenous voice which is the most compelling here, as it is the voice that, until recently, has been the least heard in academic discourses.

There are also a number of Māori legal scholars in Aotearoa New Zealand, such as Nin Tomas, Ani Mikaere, Jacinta Ruru, Carwyn Jones, Linda Te Aho, Robert Joseph, Claire Charters, Andrew Erueti, and Valmaine Toki, who are creating a valuable platform for the development of Indigenous legal theories in that country.[73]

In Australia, Hannah McGlade and Nicole Watson in 2005 both touched on theory, and in particular the role of CRT in reflecting on their own experiences as Indigenous legal academics.[74] Their work invites a more fulsome theoretical analysis of how academic institutions have generally failed to respond to the needs of Indigenous people in the law, and to expand this further, how legal institutions and the profession generally have been complicit in perpetuating the marginalisation of Indigenous voices in law. In their discourse McGlade and Watson are utilising what Christie would categorise as a mainstream theory[75] to do this rather than developing a distinctly Indigenous theory.

72 These include in Canada, for example, Brian Slattery, Patrick Macklem, Kent McNeil, Bradford Morse and Dwight Newman.

73 See, eg, Nin Tomas, 'Māori Concepts of Rangatiratanga, Kaitiakitanga, the Environment, and Property Rights' in David Grinlinton and Prue Taylor, *Property Rights and Sustainability: The Evolution of Property Rights to Meet Ecological Challenges* (Martinus Nijhoff, 2011) 219; Ani Mikaere, 'Tikanga as the First Law of Aotearoa' (2007) 10 *Yearbook of New Zealand Jurisprudence* 24; Carwyn Jones, 'A Māori Constitutional Tradition' (2014) 12(1) *New Zealand Journal of Public and International Law* 187; Jacinta Ruru, 'The Māori Encounter with Aotearoa New Zealand's Legal System' in Benjamin J Richardson, Shin Imai and Kent McNeil (eds), *Indigenous Peoples and the Law: Comparative and Critical Perspectives* (Hart, 2009) 111; Linda Te Aho, 'Indigenous Aspirations and Ecological Integrity: Restoring and Protecting the Health and Wellbeing of an Ancestral River for Future Generations in Aotearoa New Zealand' in Laura Westra, Klaus Bosselmann and Colin Suskolne (eds), *Globalisation and Ecological Integrity in Science and International Law* (Cambridge Scholars, 2011) 346; Robert Joseph, 'Reclaiming Legal Space for the First Law of Aotearoa-New Zealand' (2009) 17 *Waikato Law Review* 74; Claire Charters, 'A Self-Determination Approach to Justifying Indigenous Peoples' Participation in International Law and Policy Making' (2010) 17(2) *International Journal on Minority and Group Rights* 215; Andrew Erueti, 'The Recognition of Indigenous Peoples' Rights to Traditional Lands: The Evaluation of States by International Treaty Bodies' in Claire Charters and Andrew Erueti, *Māori Property Rights and the Foreshore and Seabed: The Last Frontier* (Victoria University Press, 2007) 175; Valmaine Toki, 'Tikanga Māori – A Constitutional Right? A Case Study' (2014) 40(1) *Commonwealth Law Bulletin* 32. For an example of non-legal Indigenous theorising, see Linda Tuhiwai Smith, *Decolonising Methodologies: Research and Indigenous Peoples* (Zed Books, 2nd edn, 2012).

74 Hannah McGlade, 'The Day of the Minstrel Show' (1995) 6(8) *Indigenous Law Bulletin* 16; Nicole Watson, 'Indigenous People in Legal Education: Staring into a Mirror without Reflection' (1995) 6(8) *Indigenous Law Bulletin* 4.

75 Christie, above n 62, 213.

More recently, Christine Black and Irene Watson have moved the discussion forward with their respective publications. Black, for example, writes that her book

> is meant to open up an innovative theoretical framework that draws on Australia's ancient indigenous jurisprudence from Australia. … I wanted the book to be the voice of the Indigenous and the way in which they see their Law, rather than the mediated voices of the non-Indigenous, on one hand, and the Indigenous voices that connote our legal tradition through a Western legal lens, on the other. The book is offered as an alternative legal reading for Indigenous peoples to explore and consider their own culture's legal theory.[76]

Watson has expressed similar sentiments in her book.[77]

Apart from Black and Watson, both of whom have started to develop Indigenous legal theories, Indigenous legal scholars in Australia have generally not yet fully engaged with the idea of developing Indigenous legal theories, unlike their Canadian (and perhaps their Aotearoa New Zealand) counterparts.[78]

Recognition of Indigenous title to land appears to have been an important factor in creating a space for Indigenous people to start theorising about the nature of their rights.[79] It took a while for Indigenous legal academics in Canada to move into this space following the *Calder* decision in 1973,[80] and there appears to be a similar time lag occurring in Australia following *Mabo v Queensland [No 2]* ('*Mabo*'),[81] almost 20 years later. This is not a criticism of Indigenous legal academics in either country by any means, because at the time of these decisions there were so very few (if any) Indigenous academics in the legal academy who could stamp their mark on the debate.[82]

76 C L Black, *The Land is the Source of the Law: A Dialogic Encounter with Indigenous Jurisprudence* (Routledge, 2011) 11–12.

77 Irene Watson, *Aboriginal Peoples, Colonialism and International Law* (Routledge, 2015) 5.

78 Virginia Marshall, although not taking a theoretical approach, has started the conversation in relation to Aboriginal water rights: Virginia Marshall, *Overturning Aqua Nullius: Securing Aboriginal Water Rights* (Aboriginal Studies Press, 2017).

79 Dale Turner, *This Is Not a Peace Pipe: Towards a Critical Indigenous Philosophy* (University of Toronto Press, 2006) 90–1.

80 *Calder v Attorney-General of British Columbia* [1973] SCR 313. For an in-depth look at the impact of this case, see Hamar Foster, Jeremy Webber and Heather Raven (eds), *Let Right Be Done: Aboriginal Title, the Calder Case and the Future of Indigenous Rights* (UBC Press, 2007).

81 (1992) 175 CLR 1.

82 The Indigenous legal academics in Canada mentioned above were all too young to have been writing in the 1970s following the *Calder* decision. Indeed, a number of them were pioneers as Indigenous legal academics: see, eg, Patricia Monture-Angus, *Journeying Forward: Dreaming Aboriginal Peoples' Independence* (Pluto Press, 2000) 18, n 1.

In Australia, the first Indigenous law graduate was Mick Dodson in 1974.[83] Australia has not yet reached a critical mass of Indigenous legal academics with the luxury of being able to focus on matters of pure theory. Like many Indigenous role models, they can be overburdened with responsibilities to the Indigenous cause (and their own communities), including calls to direct their attention to practical research. They may also consider theorising to be an inherently Western discourse and therefore antithetical to their academic aspirations.[84]

Christie has penned his thoughts on the need for developing Indigenous legal theories, and the general forms they could take.[85] In doing so, he highlights various dangers facing the Indigenous legal scholar. One is the extent to which Indigenous legal scholars should draw upon non-Indigenous theories, 'when the suspicion is that in various forms and guises these very theories have historically played a major role in the marginalisation and colonisation of Indigenous peoples'.[86] Although not averse to drawing upon non-Indigenous theories, his view is that Indigenous theorists who do so should 'exercise extreme caution and be highly critical';[87] they should not simply adopt a theory without close consideration of the context in which it evolved and by whose version or notion of 'truth' it was informed.[88] Developing Indigenous legal theories, if one considers the dangers identified by Christie, is going to be a challenging task for any Indigenous legal scholar.

Although Christie himself does not attempt to develop an Indigenous legal theory, he considers 'two examples of legal theorising at work' within Canada's legal academy, these being works by John Borrows[89] and Patricia Monture-Angus.[90] Borrows, he notes, is not overt in his theorising, but it can be extrapolated from the use he makes of various theoretical positions.[91] Monture-Angus, on the other hand, he sees as clearly attempting to address 'a serious lack of

83 By 1990, there were only 21 law graduates: Daniel Lavery, 'The Participation of Indigenous Australians in Legal Education' (1993) 4(1) *Legal Education Review* 177, 179. By 2000, there were an additional 118 Indigenous law graduates: Heather Douglas, 'The Participation of Indigenous Australians in Legal Education 1991–2000' (2001) 24 *University of New South Wales Law Journal* 485, 488. By 2009, there were approximately 500 Aboriginal and Torres Strait Islander law graduates: Phillip Rodgers-Falk, *Growing the Number of Aboriginal and Torres Strait Islander Law Graduates: Barriers to the Profession* (September, 2011) <https://docs.education.gov.au/system/files/doc/other/rodgers-falk_2011.pdf>. This, however, does not tell us where those graduates ended up. Insofar as any became legal academics, in 2000 there were very few, and those were usually in casual, part-time or lower-level academic positions: Douglas, above n 83, 486–8.

84 Christie, above n 62, 211–3.

85 Ibid 202–10.

86 Ibid 213.

87 Ibid 196.

88 Ibid 215–18.

89 John Borrows, *Recovering Canada: The Resurgence of Indigenous Law* (University of Toronto Press, 2002) ch 5.

90 Monture-Angus, *Journeying Forward*, above n 82.

91 Christie, above n 62, 223–4.

theoretical underpinning for the vast sea-changes in Canadian constitutional law concerning Aboriginal rights that have occurred over the last few decades'.[92]

Reading Borrows and Monture-Angus, one is struck by the personal flavour of their work, for they are unashamedly personal in their narratives.[93] Both scholars also emphasise that their work is intended for an Indigenous audience.[94]

But it is perhaps the non-Indigenous audience that is just as much in need of being exposed to writing (and theorising) as the Indigenous audience to which these works are primarily directed. If non-Indigenous scholars are to move towards a more comprehensive understanding of Indigenous perspectives, then they need to hear the Indigenous voice, being a voice that is not simply using non-Indigenous language and concepts to convey its message.[95]

A desire to write in a less conventional and more Indigenous style is something about which Australian Indigenous academic Loretta Kelly has expressed strong views:

> I am, at this moment, putting the finishing touches to a textbook. Needless to say, I'm quite fed-up with legal jargon and academic speak. Story-telling is a huge part of my traditional culture so I'll take this opportunity to tell a bit of my story as a legal academic – even though writing 'in the narrative' is not so acceptable in legal scholarship. I'm choosing to speak from my heart – 'to inject narrative, perspective, and feeling'.[96]

Emerging from the theoretical writings of the Indigenous Canadian scholars such as Borrows and Monture-Angus and the more practical focus of the writings of Australian Indigenous legal scholars is the need to change the way we think about the Eurocentrism/colonialism inherent in the dominant legal system. Consequently, we need to consider the nature of our legal institutions and our law schools, and how they engage with Indigenous students and issues. A crucial first step is for Indigenous scholars to become what Dale Turner calls 'word warriors', being 'an indigenous person who has been educated in the legal and political discourses of the dominant culture'.[97] The role of the word warrior is to 'reconcile the forms of knowledge rooted in indigenous communities with the legal and political discourses of the state'.[98] This is a difficult space for Indigenous people to

92 Ibid 227.
93 Monture-Angus, *Journeying Forward*, above n 82, Introduction.
94 Ibid 17; John Borrows (Kegedonce), *Drawing Out Law: A Spirit's Guide* (University of Toronto Press, 2010) xi.
95 As a companion piece to *Drawing Out Law*, Borrows has also published a work entitled *Canada's Indigenous Constitution* (University of Toronto Press, 2010), a more conventional legal text which 'contains a more explicit discussion of Anishinabek laws in "Western" legal and philosophical terms'. Borrows, *Drawing Out Law*, above n 94, xiv.
96 Loretta Kelly, 'A Personal Reflection on Being a Law Academic' (2005) 6(8) *Indigenous Law Bulletin* 19, 19 [citations omitted].
97 Turner, above n 79, 119.
98 Ibid 7.

inhabit. Monture-Angus writes of her personal difficulties as an Indigenous female academic, leading her to relinquish her position in the law school at a prestigious university.[99] The difficulties of being an Indigenous legal academic operating in a legal system that is based in a culture that is not one's own (and has been the source of dispossession and oppression) is something which is not lost on Indigenous academics in Australia.[100]

Taylor goes on to argue that word warriors are only part of the picture. There needs to be an intellectual division of labour between word warriors and what he terms '"indigenous philosophers" – indigenous intellectuals who possess the privileged forms of indigenous knowledge'.[101] These are elders – respected senior people who are recognised by their communities as being holders of Indigenous knowledge. Word warriors cannot do it alone, they need to be guided by their elders; they must 'listen to their "indigenous philosophers" while engaging with the intellectual and political practices of the dominant culture'[102] otherwise their efforts at the interface of Western and Indigenous law may be seen to lack legitimacy. Watson has expressed a similar view in relation to the role of Indigenous legal academics in Australia, articulating that they need to be guided by their community and in particular those who have a 'proper and rightful voice to speak for the country'.[103]

What, then, is the role for a non-Indigenous legal scholar like me? This is one of the many questions posed by Watson.[104] Kelly argues for non-Indigenous academics to vacate this space,[105] contending that the vacuum created by non-Indigenous academics ceasing to write about Aboriginal people 'will draw more Aboriginal researchers and practitioners to the publisher's table'.[106] Christie, on the other hand, is of the view that there can be a role for non-Indigenous scholars, stating that '[i]t is certainly open to non-Indigenous scholars, for example, to study how the legal system of the dominant society impacted (and continues to impact) on the lives of Indigenous peoples'.[107] Taylor, too, sees a role for non-Indigenous scholars, noting that '[t]here are many non-Indigenous intellectuals who can help indigenous peoples make their arguments count'.[108]

Clearly non-Indigenous scholars cannot speak for Indigenous people, nor can they develop Indigenous legal theories. They can, however, take on the roles envisaged by Christie and Taylor and use their training in the dominant Western

99 Monture-Angus, *Journeying Forward*, above n 82, 9.
100 See, eg, 'Special Focus Edition: Racism in Legal Education' (2005) 6(8) *Indigenous Law Bulletin*.
101 Turner, above n 79, 72.
102 Ibid 8. See also ch 5, 119–20.
103 Irene Watson, 'Some Reflections on Teaching Law: Whose Law, Yours or Mine?' (2005) 6(8) *Indigenous Law Bulletin* 23, 24.
104 Ibid.
105 Although co-authorship or joint research would appear to be acceptable: Kelly, above n 96, 21.
106 Ibid 20.
107 Christie, above n 62, 209.
108 Turner, above n 79, 120.

legal system under which they operate, informed by practical experience working with Indigenous communities, to expose how the dominant Western legal system has marginalised or silenced (as the case may be) Indigenous voices, and suggest how the law might be used, or amended, to empower those voices. The empowering of Indigenous voices in this way may be a step towards the reconceptualisation of the role of Indigenous law, at least in relation to natural resource and water management issues.

It could be argued that this is merely perpetuating the dominant legal system. That is, merely suggesting changes to the law to accommodate Indigenous interests does not actually change the underlying Eurocentric/colonial biases in the legal system. However, by exposing the deep imperfections in our current legal system and by suggesting changes that will give a meaningful voice within that system to Indigenous people, the non-Indigenous legal scholar can seek to change that underlying bias. That, to a large extent, is the purpose of this book.

Valuing the Indigenous voice

There are other reasons why a legislatively recognised space for Indigenous voices is important. There is symbolic value for Indigenous people in having representation (and consequently their knowledge) specifically acknowledged as being necessary. The importance of symbolism can often be underestimated, and is arguably 'the most important, and most undervalued, function of legislation'.[109] Legislative recognition accords respect to Indigenous knowledge and cultural values. Symbolic recognition, however, needs to be accompanied by legislative provisions giving substantive effect to that recognition. The reforms proposed in this book combine both symbolic recognition and the practical application of that recognition.

There is also educative value in having Indigenous representation specifically acknowledged. Government authorities may well be aware of the value of Indigenous knowledge in decision-making.[110] However, stakeholders and the wider community may be less aware of its value, or dismissive of its relevance. This may be of particular importance where the Indigenous population is much smaller and therefore less visible than elsewhere.[111]

109 Matthew Palmer, 'The Treaty of Waitangi in Legislation' (2001) *New Zealand Law Journal* 207, 209.

110 In Australia, see, eg, *West Gippsland Waterway Strategy 2014–2022* (West Gippsland Catchment Management Authority, 2014) ch 7; *Glenelg Hopkins Waterway Strategy 2014–2022* (Glenelg Hopkins Catchment Management Authority, 2014) pt 1.4.

111 For example, in Victoria, Indigenous people comprise approximately 0.7 per cent of the population, the lowest proportion of any Australian state: Australian Bureau of Statistics, *2076.0 – Census of Population and Housing: Characteristics of Aboriginal and Torres Strait Islander Australians, 2011 – Population Distribution and Structure* (16 February 2018) <www.abs.gov.au/ausstats/abs@.nsf/Lookup/2076.0main+features1102011>.

Legislative recognition is also of practical value in that Indigenous people will have a legislative basis to justify their participation, which is a stronger and more secure basis upon which to rely than policies or strategies. It should be standard practice to have Indigenous views represented in water management decisions, whether through consultation processes or Indigenous representation on decision-making authorities. Legislative recognition would promote this, to the point where it is no longer questioned.

Finally, and most importantly, it is something that Indigenous people themselves have been calling for, particularly in Australia.[112]

Legislative recognition will not meet all Indigenous aspirations for water management,[113] and Indigenous people will continue to assert their inherent water rights irrespective of legal recognition by the dominant Western legal system. Indeed, there are many examples of Indigenous-led initiatives occurring outside of legislative recognition in both Australasia and North America. The Gunditjmara in western Victoria, for example, have been working for many years to have the Budj Bim Cultural Landscape, which contains one of the largest and oldest aquaculture systems in the world, included on the World Heritage List.[114] The Ngarindjerri in South Australia have developed a management plan with the state government over the Sugar Shack wetland complex.[115] And in Canada, an initiative by the five Mi'kmaq Chiefs of Unama'ki in 2003 resulted in the Bras d'Or Lakes Collaborative Environmental Planning Initiative, to improve the health of the Lakes.[116]

Nonetheless, legislative recognition is still an important tool for effecting change, underpinning and strengthening the Indigenous voice in the water management landscape.

Concluding remarks

The justifications for seeking a strong and effective Indigenous voice in water management are numerous and compelling. Including Indigenous perspectives is vital for the sustainable management of water, now and into the future.

112 See, eg, *FVTOC Water Policy Statement 2014*, above n 56, 8, 13; North Australian Indigenous Land and Sea Management Alliance, *A Policy Statement on North Australian Indigenous Water Rights* (November 2009) 3; *Mary River Statement* (6 August 2009).

113 See, eg, Alex Streensa, 'Accommodating Indigenous Cultural Values in Water Resource Management: The Waikato River, New Zealand; the Murray-Darling Basin, Australia; and the Colorado River, USA' (Paper presented at the Australian Agricultural & Resource Economics Society's Annual Conference, Cairns, 11–13 February 2009) 10.

114 See, eg, <http://whc.unesco.org/en/tentativelists/6167>.

115 Steve Hemming et al, 'A New Direction for Water Management? Indigenous Nation Building as a Strategy for River Health' (2017) 22(2) *Ecology and Society* 13.

116 See, eg, <https://brasdorcepi.ca/>. See also Heather Castleden et al, 'Reconciliation and Relationality in Water Research and Management in Canada: Implementing Indigenous Ontologies, Epistemologies, and Methodologies', in Steven Renzetti and Diane P Dupont (eds), *Water Policy and Governance in Canada* (Springer, 2017) 85.

Indigenous people need effective tools by which they can participate meaningfully in water management. The fundamental proposition of this book is that legal recognition, in conjunction with initiatives such as those outlined above, and the right policy context and political environment, is one such tool.

Part B

Case study

Australia – historical context

2 History of water law in Australia

The marginalisation of Indigenous perspectives in water management is a product of the Eurocentric legal system that Australia inherited on colonisation, as evolved over time. Accordingly, that system should be reoriented to be more inclusive of Indigenous views. It is therefore fitting that those of us with an understanding of how that system operates seek ways to change it to better accommodate Indigenous views. However, to ascertain what changes might be appropriate, we need to know what the legal framework currently provides, and how it was developed. It is those matters to which this book now turns.

Common law water rights

An evaluation of the development of water law and regulation in Australia would not be complete without an understanding of the historical background and the role of the common law, given that 'the rules of the common law remain the foundation of Australian water law'.[1] But first a fundamental point of clarification. With thousands of years of experience of managing water resources prior to colonisation, Australia's Indigenous peoples were the first water managers. It is the non-Indigenous history of water law that this chapter will outline, but it is imperative in the context of this book to acknowledge that there were pre-existing Indigenous water management regimes in Australia prior to the introduction of common law water rights.

Various texts go into detail on the history and nature of common law rights to water,[2] so I will not replicate that material here. A summary of the main elements will suffice, commencing with the reception of the common law into Australia.

1 D E Fisher, *Water Law* (LBC Information Services, 2000) 64.

2 Ibid, ch 4; Alex Gardner, Richard Bartlett and Janice Gray, *Water Resources Law* (LexisNexis Butterworths, 2nd edn, 2018) ch 8; Kate Stoeckel, *Australian Water Law* (Thomson Reuters, 2012) ch 2.

Reception of the common law into Australia

It was on the assertion of sovereignty by the British Crown that Australia inherited British common law. Justice Brennan in *Mabo* explained it as follows:

> [T]he theory which underpins the application of English law to the colony of New South Wales is that English settlers brought with them the law of England and that, as the indigenous inhabitants were regarded as barbarous or unsettled and without a settled law, the law of England, including the common law became the law of the Colony (so far as it was locally applicable) as though NSW were 'an uninhabited country ... discovered and planted by English subjects'.[3]

The concept of the need for local applicability in the above passage stems initially from Blackstone, who noted that the colonisers brought with them only so much of English law that was applicable to 'their own situation and the condition of [the] infant colony'.[4] Despite the existence of this concept, it appears to have been somewhat overlooked, particularly in relation to Indigenous people, the existence of whom had little effect on the development of the common law in Australia until *Mabo*. It would also appear to have been overlooked in relation to the common law regarding water.

That the common law inherited by Australia included the common law rules relating to water was made clear by Windeyer J in *Gartner v Kidman* where he stated that 'it is beyond doubt that these rules [relating to water] are a part, and an important part, of the common law that Australia has inherited'.[5]

What, then, is the nature of these common law rules which Australia inherited? Common law rules relating to water developed out of the concept that water in its natural state is *publici juris*, or *res communes*, common property.[6] Thus there could be no absolute ownership of water.

Although the common law did not recognise absolute ownership of water, it did recognise a right of access to water to those through whose land the water flowed. These rights of access were known as riparian rights, the principles of which were set out in the classic case of *Embrey v Owen*. [7] The uses to which those

3 (1992) 175 CLR 137–8 [citations omitted].

4 William Blackstone, *Commentaries on the Laws of England* (Clarendon Press, 2nd edn, 1766) Vol 1, 107.

5 (1962) 108 CLR 12, 23.

6 Clark and Renard note that 'whilst Blackstone spoke of *res communes*, a number of early cases classified running water as *publici juris*', Sandford D Clark and Ian A Renard, *The Law of Allocation of Water for Private Use, Volume 1: The Framework of Australian Water Legislation and Private Rights* (Research Project 69/16, Australian Water Resources Council, 1972) 52. See also Poh-Ling Tan, 'The Changing Concepts of Property in Surface Water Resources in Australia' (2002) 13(4) *Journal of Water Law* 269, 270–2.

7 6 Ex 353, 369; 155 ER 579, 585–6 (Parke B).

riparian rights could be put were later summarised in *Attwood v Llay Main Collieries* as follows:

> A riparian owner may take and use the water for ordinary purposes connected with the riparian tenement (such as domestic purposes or the wants of his cattle), and that in the exercise of his right, he may exhaust the water altogether; that he may also take and use the water for extraordinary purposes, if such user be reasonable and be connected with the riparian tenement, provided that he restores the water so taken and used substantially undiminished in volume and unaltered in character; and lastly, that he has no right whatever to take the water and use it for purposes unconnected with the riparian tenement.[8]

Riparian rights only applied to a natural watercourse, not to an artificially constructed channel.[9] Furthermore, it had to be in relation to water flowing through a defined channel and not to water flowing indiscriminately over the surface of the land.[10]

The common law rule in relation to water flowing indiscriminately over the surface of the land has been described as the 'rule of capture'.[11] This rule was based on the fact that the flow of water was both uncertain and unpredictable. Unlike water flowing through a natural watercourse, whereby a landowner's rights were subject to the rights of neighbouring landowners along the watercourse, the rule of capture had no such limitation. It entitled a landowner to appropriate all the water on the property and 'to get rid of it in any way he pleases'.[12]

Groundwater, like surface water, was also seen to be of two types: water flowing in a defined channel and water flowing indiscriminately – that is, percolating through the soil. Where underground water could be established as flowing through a defined channel, the riparian rights relating to natural watercourses applied.[13] In relation to groundwater percolating through the soil, the common law 'rule of capture' applied.[14]

These principles arising from the British cases were cited and affirmed in early Australian cases,[15] and continue to be cited as reflecting the common law

8 [1926] Ch 444, 458 (Lawrence J).
9 *Gartner v Kidman* (1962) 108 CLR 12, 24 (Windeyer J, Dixon CJ concurring).
10 *Rawstron v Taylor* (1855) 11 Ex 369; 156 ER 873, 880; *Broadbent v Ramsbotham* (1856) 11 Ex 602, 305; 156 ER 971, 976.
11 Gardner, Bartlett and Gray, above n 2, 192. The concept was first established in *Acton v Blundell* (1843) 152 ER 1223, 1235 in relation to groundwater, and subsequently established in relation to surface water flowing indiscriminately in *Rawstron v Taylor* (1855) 156 ER 873; 11 Ex 369; and *Broadbent v Ramsbotham* (1856) 156 ER 971; 11 Ex 602.
12 *Rawstron v Taylor* (1855) 11 Ex 873, 382; 156 ER 873, 880 (Parke B).
13 Gardner, Bartlett and Gray, above n 2, 189; *Chasemore v Richards* (1859) 7 HL Cas 349, 374; 11 ER 140, 150 (Lord Chelmsford).
14 Gardner, Bartlett and Gray, above n 2, 188.
15 *Dunn v Collins* (1867) 1 SALR 143; *Perth Corporation v Halle* (1911) 13 CLR 393; *Jones v Kingborough* (1950) 82 CLR 282.

principles inherited by Australia, most recently by the High Court in *ICM Agriculture v Commonwealth*.[16]

Common law rules in the Australian context

Australia experienced many difficulties with the application of these common law rules, all of which have been well documented.[17] As Stephen J commented in *Hanson v Grassy Gully Gold Mining Co*:

> It cannot be denied that for years and years past the question of the rights of riparian owners in this country, where the conditions are so totally different from the condition of things in England, has been a source of almost insuperable difficulty.[18]

The riparian regime emerged 'in the context of the significant number of rivers and streams crisscrossing England, possessed of a temperate climate, such that there would not be many parts of the country at a great distance from a surface water supply'.[19] By contrast, Australia is widely regarded as the driest inhabited continent on earth,[20] with extreme variability in rainfall and runoff both spatially and temporally.[21]

In nineteenth-century England, water resources were relatively abundant, and thus the common law 'served a well-watered intensively developed England quite well'.[22] However, in relation to Australia, '[t]he continued expansion of the various colonies in the second half of the 19th century, coupled with a series of droughts in the 1880s, underscored the inadequacy of the common law foundation for water law'.[23]

A further difficulty of the common law relates to the distinction between underground and surface water. Unlike surface water, water percolating underground was 'a process of nature not apparent', requiring 'the evidence of scientific men'[24] to resolve disputes. With scientific advances over the years, it is now clear

16 (2009) 240 CLR 140.

17 Gardner, Bartlett and Gray, above n 2, 199–203; Edwyna Harris, 'Colonialism and Long-run Growth in Australia: An Examination of Institutional Change in Victoria's Water Sector during the Nineteenth Century' (2008) 48(3) *Australian Economic History Review* 266, 272–275; Sandford Clark and Ian Renard, 'The Riparian Doctrine and Australian Legislation' (1969–1970) 7 *Melbourne University Law Review* 475, 477–9.

18 (1900) 21 NSWLR 271, 275.

19 Gardner, Bartlett and Gray, above n 2, 176.

20 David Ingle Smith, *Water in Australia: Resources and Management* (Oxford University Press, 1998) 4.

21 Ibid 4–16.

22 Lee Godden, 'Water Law Reform in Australia and South Africa: Sustainability, Efficiency and Social Justice' (2005) 17(2) *Journal of Environmental Law* 181, 186.

23 Ibid 186–7.

24 *Chasemore v Richards* (1859) 7 HL Cas 349, 379; 11 ER 140, 152 (Lord Cranworth).

that surface water and groundwater are interconnected, and therefore the distinction between them and the associated common law rules are at odds with modern understandings of water resources. Yet, despite the increased understanding of the interconnectedness between surface water and groundwater, it is only relatively recently that the regulatory regimes in Australian jurisdictions have begun to catch up with this understanding, with the distinction persisting to some extent today.[25] The distinction is also at odds with the Indigenous world view of water and indeed of country, which is much more holistic in its approach.[26]

The Victorian response

Victoria has been credited as being at the forefront of the development of modern water management in Australia,[27] whose 'development has been typical, though perhaps more dramatic'[28] of most of the states.

The need for a management regime was not originally apparent in the early years of the colony of Victoria as there was little pressure on water resources. This was largely due to a small population and relatively low-impact land use, namely grazing.[29] Therefore the common law rules and riparian rights in particular 'remained relatively stable'.[30] However, after the discovery of gold in the 1850s there was significant growth in the Victorian population. Between 1851 and 1861, the population increased by around 600 per cent,[31] much of this in the goldfields.[32] Mining required the use of large amounts of water,[33] leading to an increase in demand, particularly in the goldfields regions. The relative scarcity of

25 Under the *Water Act 1989* (Vic), water supply protection area plans can be developed for surface water, groundwater or both. To date, only one such plan has been developed for both surface and groundwater: *Upper Ovens River Water Supply Protection Area: Water Management Plan* (11 January 2012).

26 See, eg, Sarah Yu, *Ngapa Kunangkul: Living Water. Report on the Aboriginal Cultural Values of Groundwater in the La Grange Sub-basin* (University of Western Australia, 2nd edn, May 2000); Sandy Toussaint, Patrick Sullivan and Sarah Yu, 'Water Ways in Aboriginal Australia: An Interconnected Analysis' (2005) 15(1) *Anthropological Forum* 61; Marcia Langton, 'Earth, Wind, Fire and Water: The Social and Spiritual Construction of Water in Aboriginal Societies' in Bruno David, Bryce Barker and Ian J McNiven (eds), *The Social Archaeology of Australian Indigenous Societies* (Aboriginal Studies Press, 2006).

27 J M Powell, *Watering the Garden State: Water, Land and Community in Victoria 1834–1988* (Allen & Unwin, 1989) 7; Clark and Renard, *The Law of Allocation of Water for Private Use*, above n 6, 141; Fisher, above n 1, 5.

28 Clark and Renard, *The Law of Allocation of Water for Private Use*, above n 6, 141.

29 Harris, 'Colonialism and Long-run Growth in Australia', above n 17, 273.

30 Ibid.

31 William Henry Archer, Registrar General, *Census of Victoria, 1861. Population Tables – Part 1 – Inhabitants and Houses: Showing the Numbers and Distribution of the People* (Melbourne, 25 April 1862) xiii.

32 Ibid xiv; Powell, *Watering the Garden State*, above n 27, 52.

33 Peter Davies, Susan Lawrence and Jodi Turnbull, 'Harvesting Water on a Victorian Colonial Goldfield' (2011) 29 *Australasian Historical Archaeology* 24, 24.

water resources led to much conflict.[34] Accordingly, various statutes were enacted to deal with these issues.[35] Although they did not always achieve their goals,[36] these legislative measures were 'an early move towards the principle of public control and ownership of a crucial resource, establishing an important precedent for future management'.[37]

In the 1860s the gold began to run out, leading to an influx of miners into the agricultural sector.[38] Following land reform in the 1860s, precipitated by this sudden increase in population and subsequent decline of mining, land use intensified, changing from largely grazing to crop farming which was heavily reliant on irrigation. Common law riparian rights became less effective 'because it [crop farming] required diversions away from the river or stream'.[39] This intensified land use increased demand for water, along with the irrigation infrastructure to support it.

A severe drought from 1871 to 1881 led to mounting political pressure on the colonial government to ensure a more stable water supply. Thus came the enactment of the *Water Conservation and Distribution Act 1881* (Vic) followed by the *Water Conservation Act Amendment Act 1883* (Vic), with that Act being further amended in 1885. The following year, the government enacted the seminal *Irrigation Act 1886* (Vic) ('*Irrigation Act*').

Alfred Deakin, later to become the Prime Minister of a newly federated Australia, was a key mover behind these developments. In December 1884, a Royal Commission on Water Supply was established, and Deakin, who was at that time Victoria's Solicitor-General,[40] was appointed Chair. Deakin took a keen interest in developments in the United States, particularly in western USA which he visited shortly after his appointment. Among the states that he visited were Arizona, California and Colorado, all of which were known for their irrigation works.[41] The findings by Deakin in the Royal Commission reports formed the basis for the *Irrigation Act.*[42]

34 Ibid, 31; Peter Davies and Susan Lawrence, 'A Mere Thread of Land: Water Races, Gold Mining and Water Law in Colonial Victoria' (2014) 16 *Journal of Australian Colonial History* 165, 165.
35 Davies and Lawrence, above n 34, 176–84.
36 Powell, *Watering the Garden State*, above n 27, 50.
37 Davies and Lawrence, above n 34, 183.
38 Powell, *Watering the Garden State*, above n 27, 48, 62; Clark and Renard, *The Law of Allocation of Water for Private Use*, above 6, 156; Edwyna Harris, 'Lobbying for Legislation: An Examination of Water Rights Transition in Colonial Victoria, Australia 1840–1886' (Discussion Paper 12/06, Department of Economics, Monash University, 2006) 7–8.
39 Harris, 'Colonialism and Long-run Growth in Australia', above n 17, 274.
40 Victoria, *Parliamentary Debates, Session 1884 – Legislative Assembly and Legislative Council – Vol XLV* [front matter – The Cabinet].
41 Gardner, Bartlett and Gray, above n 2, 206.
42 Clark and Renard, 'The Riparian Doctrine and Australian Legislation', above n 17, 487.

One of the most significant aspects of the *Irrigation Act* was the effect it had on common law riparian rights. Deakin had wanted to abolish riparian rights,[43] but he was unable to convince parliament, which eventually compromised with the following vesting provision:[44]

> The right to the use of all water at any time in any river stream watercourse lake lagoon swamp or marsh shall for the purposes of this Act in every case be deemed to be vested in the Crown until the contrary be proved by establishing any other right than that of the Crown to use such water …[45]

The same provision also provided for

> the general right of all persons to use water for domestic and stock supply from any stream watercourse lake lagoon swamp or marsh vested in the Crown, and to which there is access by a public road or reserve.

Elements of this latter part of the provision can still be found today in s 8(1) of the *Water Act 1989* (Vic) ('*Water Act 1989*'). By vesting the right to the use of all water in the Crown, the *Irrigation Act* essentially 'prevented the establishment of further riparian rights and effectively nationalised Victoria's water resources'.[46] This vesting in the Crown was to become a feature of water legislation around Australia, with New South Wales being the first to follow suit. However, New South Wales took it a step further – the *Water Rights Act 1896* (NSW) included a provision which vested not only the 'use' but the 'flow and control' of water as well. This terminology was later adopted in Victoria's *Water Act 1905* (Vic) ('*Water Act 1905*'), and by both Queensland and Western Australia.[47]

The other significant aspect of the *Irrigation Act* was that it tied water allocations to grants of land,[48] a nexus which 'remained unbroken for over a hundred years'.[49] For Aboriginal Victorians who had been dispossessed of their land, the effect of this was to shut them out of those initial water allocations. This had significant implications for when that nexus was eventually broken.

Another limitation on the continuation of riparian rights was that unlike in other jurisdictions whereby grants of land next to a river were *ad medium filum aquae* ('to the centre thread of the water'), it was the Victorian practice

43 Gardner, Bartlett and Gray, above n 2, 206.

44 Ibid, 207.

45 *Irrigation Act* s 4.

46 J M Powell, *Environmental Management in Australia, 1788–1914: Guardians, Improvers and Profit: An Introductory Survey* (Oxford University Press, 1976) 132.

47 Clark and Renard, 'The Riparian Doctrine and Australian Legislation', above n 17, 489; Gardner, Bartlett and Gray, above n 2, 185.

48 *Irrigation Act* s 139. See also Godden, 'Water Law Reform in Australia and South Africa', above n 22, 187; Edwyna Harris, 'Institutional Change and Economic Growth: The Evolution of Water Rights in Victoria, Australia 1850–1886' (2007) 26 (2) *Economic Papers* 118, 123.

49 Godden, 'Water Law Reform in Australia and South Africa', above n 22, 187.

(introduced sometime in the mid-1860s) 'to reserve a strip of land adjacent to the river over which a landowner could not obtain title'.[50] The consequence of this was effectively to deny a landowner riparian rights.[51] However, the practice 'operated only in relation to major rivers and boundary streams',[52] and therefore landowners with streams running through their allotments would not have been affected. This practice was later enshrined in the *Water Act 1905.* [53]

Importantly, it was the 1905 Act which arguably abolished surface water common law riparian rights.[54] As noted above, the vesting provision in the *Irrigation Act* was amended to reflect the wording in the *Water Rights Act 1896* (NSW), which provided for 'the right to the use and flow and to the control of the water'. It was this new vesting provision, contained in s 4 of the *Water Act 1905*, in combination with both s 5, which declared that the beds and banks of boundary rivers remained the property of the Crown, and s 6, which prohibited the taking of water except in accordance with the relevant Act, that arguably abolished common law riparian rights.[55]

The *Water Act 1905* also established the State Rivers and Water Supply Commission, the purpose of which was 'to provide overall state control and administration of water development',[56] thus cementing the principle of public control of a crucial resource, foreshadowed in the early mining legislation.

The *Irrigation Act*, the *Water Act 1905*, and the various later amending Acts, culminating in the *Water Act 1958*, only applied to surface waters. There was no regime in place in Victoria to manage groundwater – that is, until the enactment of the *Groundwater Act 1969* (Vic) ('*Groundwater Act*'). Therefore, until 1969, while rights in surface waters were dealt with by statute, common law rights to groundwater still applied.

The *Groundwater Act* contained a vesting provision, s 47. This provision was based on s 4 of the *Water Act 1958* (Vic), which had found its genesis in s 4 of the *Irrigation Act.*

It was explained by the Minister that the main reason for this was the interconnectedness of surface water and groundwater and the consequences of not treating them the same way.[57] The Minister accordingly 'viewed it as having the

50 Clark and Renard, 'The Riparian Doctrine and Australian Legislation', above n 17, 483–4.
51 Ibid 484.
52 Ibid 492.
53 *Water Act 1905* s 5. See also Poh-Ling Tan, 'Legal Issues Relating to Water Use' in Catherine Mobbes and Ken Moore, *Property: Rights and Responsibilities – Current Australian Thinking* (Land and Water Australia, 2002) 16. Note that this meant that there was public land adjacent to and under rivers, over which native title could then be recognised.
54 Clark and Renard, 'The Riparian Doctrine and Australian Legislation', above n 17, 491–2.
55 Ibid 492.
56 Gardner, Bartlett and Gray, above n 2, 208.
57 Victoria, *Parliamentary Debates*, Legislative Assembly, 23 October 1968, 1242 (W A Borthwick, Minister of Water Supply) 1243.

same effect as section 4 in the *Water Act 1958*'.[58] Thus the abolition of all common law rights to water was almost complete.

Despite the acknowledgement of the interconnectedness of surface and groundwater, it was another 30 years before a comprehensive overhaul of Victoria's water management arrangements took place, resulting in the *Water Act 1989*. It is this Act which finally abolished all common law rights to water.[59] One of the significant features of the *Water Act 1989* was the introduction of provisions enabling the transfer of water entitlements, thus commencing the process of separating water title from land.[60] For Indigenous Victorians, this would effectively amount to a double dispossession, as any subsequent recognition or return of land would not include any water allocations that had previously attached to the land.[61]

The development of the *Water Act 1989* and subsequent amendments are discussed in more detail in the next chapter.

Federation, the Australian *Constitution* and responsibility for water management

Like Victoria, each of the other states developed its own water management regime both before and after federation. This continued development at a state rather than Commonwealth level occurred because of the nature of Australia's political system, enshrined in the federal *Constitution*. In the *Constitution*, the Commonwealth has express power to make laws in respect of various specified matters, but not in relation to water management. As this factor has been instrumental in shaping Australia's current water management regime, it is appropriate to mention briefly its historical background.

Alfred Deakin, the main architect behind Victoria's *Irrigation Act* and the 'father of Australian water legislation',[62] went on to become one of ten Victorian delegates to the Constitutional Conventions of the 1890s. He, along with many of his fellow Victorians, spoke at length on issues relating to water, as did the delegates from both New South Wales and South Australia. It has been pointed out that 'the issue of the rights of States to water resources was one of the most contentious in Australia's constitutional history'.[63] During the Conventions,

58 Sandford D Clark and A J Meyers, 'Vesting and Divesting: The Victorian Groundwater Act 1969' (1969–70) 7 *Melbourne University Law Review* 237, 240.

59 *Water Act 1989* s 8(7). See Gardner, Bartlett and Gray, above n 2, 224–5.

60 Ibid s 2.

61 Tony McAvoy makes a similar point in relation to the *Water Management Act 2000* (NSW): Tony McAvoy, 'Water – Fluid Perceptions' (2006) 1(2) *Transforming Cultures eJournal* 97, 101.

62 Smith, above n 20, 150.

63 Nicholas Kelly, 'A Bridge? The Troubled History of Inter-State Water Resources and Constitutional Limitations on State Water Use' (2007) 30(3) *University of New South Wales Law Journal* 639, 641.

'monstrously long and tangled debates'[64] took place, particularly at the Adelaide and Melbourne Conventions, with the Murray River clearly taking centre stage.[65] It is not within the scope of this book to canvas these matters as other eminent scholars have already evaluated the Murray River issues and negotiation over water in the decades leading up to federation.[66] It suffices to say that it was such a controversial topic that it was unable to be resolved, resulting in only two sections relating to water being included in the *Constitution*. Those two sections are ss 98 and 100, and are of limited application.[67]

As responsibility for water management did not become one of the enumerated Commonwealth powers upon federation, the states had to rely on interstate agreements to deal with issues crossing state boundaries.[68] Many of these agreements also involved the Commonwealth, the Snowy Mountains Scheme and the various Murray River agreements being the more significant ones.

That is not to say that the Commonwealth is unable to make laws of its own with respect to water in the absence of agreement with the states. Despite limited express powers relating to water in the *Constitution*, the Commonwealth has on occasion relied on other powers to enact legislation involving water management. The Snowy Mountains Scheme, for example, was originally authorised by Commonwealth legislation by using the Commonwealth's defence power.[69] More recently, the Commonwealth had to rely on various powers to enact the *Water Act 2007* (Cth) ('*Water Act 2007*'), discussed further below.

The development of water law – federation to the present

Water management in the twentieth and twenty-first centuries consists of three distinct phases: post-federation, the reforms of the late 1970s up to 1994, and what will be referred to as the modern era, encompassing the COAG Framework Agreement of 1994, the *Intergovernmental Agreement for a National Water Initiative* ('*NWI*')[70] and the *Water Act 2007*. This last phase heralded a much more focussed national approach, albeit one which had been foreshadowed for some time.

As with much of the early history of water law in Australia, several commentators have written about these later stages of water management.[71] Thus, the

64 J A La Nauze, *The Making of the Australian Constitution* (Melbourne University Press, 1972) 153.

65 Kelly, above n 63, 639, 642.

66 Sandford D Clark, 'The River Murray Question: Part I — Colonial Days' (1971) 8 *Melbourne University Law Review* 11.

67 For more detail, see Paul Kildea and George Williams, 'The Constitution and the Management of Water in Australia's Rivers' (2010) 35 *Sydney Law Review* 595.

68 See Gardner, Bartlett and Gray, above n 2, ch 7 for a brief overview of various such interstate agreements.

69 Gardner, Bartlett and Gray, above n 2, 162.

70 *Intergovernmental Agreement on a National Water Initiative* (25 June 2004).

71 Smith, above n 20, 155–192; Gardner, Bartlett, and Gray, above n 2, ch 3; Tan, 'Legal Issues Relating to Water Use', above n 53, 16–24; in relation to Victoria, see Powell, *Watering the Garden State*, above n 27, 136–293.

following section will not seek to repeat that commentary, but merely highlight some of the main features, and, where appropriate, their relevance to Indigenous water rights and the general themes of this book.

Post-federation

The period following the introduction of the statutory vesting schemes up until the end of World War II was characterised by the construction of large state-owned dams and associated irrigation infrastructure.[72] Responsibility for the construction of these works resided with state agencies; in Victoria this was the State Rivers and Waters Commission.

One factor that fuelled the need for increased irrigation capacity during this time was the allocation of smallholdings to returned World War I soldiers, known as the soldier settler scheme. These soldier settlements were largely unsuccessful, but the impact they had was to 'increase both the size of the total area irrigated and the number of farmers dependent on irrigation'.[73]

This was also the period during which agreement on the Murray River was first reached. In 1914 following extensive negotiation, the River Murray Waters Agreement between the Commonwealth, New South Wales, Victoria and South Australia was finally agreed; however, it applied only to the Murray River rather than the much larger Murray–Darling Basin ('MDB').[74] The initial 1914 River Murray Waters Agreement arguably 'marks the beginning of serious federal government involvement in water resource planning'.[75]

Following World War II was a period of post-war reconstruction which 'herald [ed] a new era of Commonwealth involvement in resource matters'.[76] Building on the previous predilection of state governments for building large dams, this was 'the era of mega-projects founded on even larger dams, with the Commonwealth footing much of the bill'.[77] These projects were generally undertaken via grants of financial assistance to the states pursuant to the Commonwealth's power under s 96 of the *Constitution*. The Commonwealth was prepared to assist in funding these projects as they were considered of national significance and 'too large for individual states to fund'.[78] These 'mega-projects' included the Snowy Mountains

72 Smith, above n 20, 156, 162; Poh-Ling Tan et al, *Collaborative Water Planning: Context and Practice – Literature Review, Volume 1* (TRaCK Report, September 2008) 24.
73 Smith, above n 20, 158.
74 Sandford D Clark, 'The Murray-Darling Basin: Divided Power, Co-operative Solutions' (2003) 22 *Australian Resources and Energy Law Journal* 67, 70.
75 Smith, above n 20, 163.
76 Ibid 166.
77 Ibid 167. See also Lee Godden, 'Perception of Water in Australian Law: Re-examining Rights and Responsibilities' (Proceedings of the 2003 Invitation Symposium: Water – The Australian Dilemma, Academy of Technological Sciences and Engineering, Canberra, 2003) 44.
78 Smith, above n 20, 166.

Scheme, the Ord River Scheme in Western Australia, the Gordon River Scheme in Tasmania and the Burdekin Dam in Queensland.[79]

Godden has described the post–World War II era as one in which there was a 'more comprehensive viewpoint' which 'took into account water users, management purposes, and land-water connections, and the growing acceptance of the river catchment management. There was also a growing awareness of environmental and social impacts, in addition to the traditional focus on technology.'[80] However, with the possible exception of the *Groundwater Act* in Victoria, no major changes to the water management regimes occurred during this period. And not taken into account in this 'more comprehensive viewpoint' were impacts on Indigenous water needs.

The late 1970s to 1994

The impetus for reform accelerated in the late 1970s and continued throughout this period, prompted by a number of factors. The environmental movement was gaining momentum,[81] the concept of ecologically sustainable development was gaining currency, both internationally and locally,[82] and there was recognition that in many parts of Australia there were significant problems of over-allocation of water, particularly in the MDB.

The Murray River Waters Agreement of 1914 had been amended from time to time but had remained in substantially the same form until 1981. That year, in response to these issues, a 'revised Agreement was approved at a meeting of the Prime Minister and the respective Premiers'[83] and signed the following year.[84] A new agreement was subsequently reached in 1987, of note being that it now applied to the whole of the MDB, not just the Murray River.[85] This Agreement was then executed in amended form in 1992 and ratified by legislation in each of the participating jurisdictions the same year.[86] These reforms were indicative

79 See, eg, ibid 167–82 for an analysis of these four projects.

80 Godden, 'Perception of Water in Australian Law', above n 77, 45.

81 See, eg, the conflict over the Gordon River Scheme in Tasmania, culminating in the 1983 High Court Tasmanian Dams decision: *Commonwealth v Tasmania* (1983) 158 CLR 1.

82 See, eg, the United Nations Conference on the Human Environment, Stockholm, 1972; Conference Report of the World Commission on Environment and Development, *Our Common Future*, UN Doc A/42/427 (4 August 1987), the United Nations Conference on Environment and Development, Rio de Janeiro 1992, and in Australia, eg, the *Intergovernmental Agreement on the Environment* (1 May 1992); *National Strategy for Ecologically Sustainable Development* (December 1992).

83 Sandford D Clark, 'The River Murray Waters Agreement: Peace In Our Time?' (1983) 9 *Adelaide Law Journal* 108, 109.

84 'Murray Agreement Signed', *Advertiser* (Adelaide), 1 September 1982, 32.

85 Clark, 'The Murray-Darling Basin: Divided Power, Co-operative Solutions', above n 74, 71.

86 Gardner, Bartlett and Gray, above n 2, 45. This Agreement (revised in 1996) was to remain in place until 2008, when a new MDB Agreement was reached.

of the preference for cooperative federalism over Commonwealth legislative intervention.[87]

Tan notes that 'by the mid-1980s, reviews in several states had led policy makers to realise that major organisational and legislative changes were needed'.[88] It was towards the end of this period that Victoria consolidated its water management regime into the *Water Act 1989*.

In 1992, the states, territories and the Commonwealth entered into an *Intergovernmental Agreement on the Environment*, [89] accepting that 'principles of ecologically sustainable development would guide development and implementation of environmental policy and programs'.[90] That same year, the Commonwealth also finalised a *National Strategy for Ecologically Sustainable Development* ('*NSESD*'). Adopted by all levels of Australian governments, this document was seen as important in helping to set the future direction of water reform.[91] Chapter 18 of the *NSESD* relates to water resource management, noting that the challenge in this area is:

> To develop and manage in an integrated way, the quality and quantity of surface and groundwater resources, and to develop mechanisms for water resource management which aim to maintain ecological systems while meeting economic, social and community needs.[92]

The measures to achieve the objectives of water resource management were numerous, but included developing 'improved measures for effective public participation in development of water pricing and allocation policies and water resource management measures'.[93]

A further document which played an important role in water policy-making was the 1992 Industry Commission Report on Water Resources and Waste Water Disposal.[94] With its focus clearly on pricing reforms,[95] it was noted that '[t]he economic overtones of this report played a major influence on CoAG policy in

87 For an overview of cooperative federalism, see, eg, Gardner, Bartlett and Gray, above n 2, 118–19. See also Gerard Carney, *The Constitutional Systems of the Australian States and Territories* (Cambridge University Press, 2006) 11–13.

88 Tan, 'Legal Issues Relating to Water Use', above n 53, 20.

89 The text of the *Intergovernmental Agreement on the Environment* is contained in the Schedule to the *National Environment Protection Council Act 1994* (Cth).

90 Tan, 'Legal Issues Relating to Water Use', above n 53, 23.

91 John J Pigram, *Australia's Water Resources: From Use to Management* (CSIRO Publishing, revised edn, 2007) 64. See also Gardner, Bartlett and Gray, above n 2, 46.

92 *NSESD* (December 1992) 72.

93 Ibid.

94 M L Parker, T J Hundloe and D R Chapman, 'Water Resources and Waste Water Disposal' (Report No 26, Industry Commission, 17 July 1992).

95 Henning Bjørnlund, *Water Trade Policies as a Component of Environmentally, Socially and Economically Sustainable Water Use in Rural Southeastern Australia* (PhD Thesis, University of South Australia, 1999) 46.

1994'.[96] The *NSESD* specifically referred to this report as being something to which governments should have regard in order to achieve the *NSESD*'s objectives in relation to water resources management.[97]

Other concerns emerging during this period that also had a major influence on future water reform included

> environmental degradation, increasing competition for scarce and highly variable water supplies, less than optimal governance and pricing arrangements, and a realisation that opportunities to augment supplies, at least in inland areas, were limited because few cost-effective, large-scale dam sites remained in regions of high water demand.[98]

Thus the stage was set for more extensive water reform initiatives.

The modern era of water reform – 1994 to the present

The modern era of water reform commenced in 1994 with agreement by the Council of Australian Governments ('COAG')[99] on a Water Reform Framework ('1994 Framework'). This Framework was followed in 2004 by the *NWI* and in 2007 by the enactment of the *Water Act 2007*. These reforms, and particularly the legislative reforms, have been described as 'the most significant since Australian water resources statutes were first enacted over a century ago'.[100] A brief overview of each of the reforms is set out below.

The 1994 COAG Agreement on a Water Reform Framework

The issues arising during the 1970s and 1980s coalesced into two main problems of water use, namely 'the opposing demands for security for consumptive users, and the growing awareness that water needed to be allocated to ecosystem needs'.[101] This led to COAG adopting the 1994 Framework.[102] The 1994 Framework was seen 'as signalling a new urgency on the part of governments, both state and federal, to promote efficient, sustainable use of water in Australia'.[103] In terms of what it intended to achieve, the 1994 Framework stated that it

96 Tan et al, *Collaborative Water Planning*, above n 72, 27.
97 *NSESD*, above n 92, 73.
98 National Water Commission, *The National Water Initiative – Securing Australia's Water Future: 2011 Assessment* (September 2011) 4 ('*NWI Assessment 2011*').
99 COAG is the peak intergovernmental forum in Australia. For more information about COAG, see <www.coag.gov.au>.
100 Gardner, Bartlett and Gray, above n 2, xxii.
101 Tan, 'Legal Issues Relating to Water Use', above n 53, 22.
102 Council of Australian Governments, *Communiqué* (25 February 1994) Attachment A: Water Resource Policy <http://archive.coag.gov.au/coag_meeting_outcomes/1994-02-25/docs/attachment_a.cfm>.
103 Pigram, above n 91, 64.

embraces pricing reform based on the principles of consumption-based pricing and full-cost recovery, the reduction or elimination of cross-subsidies and making subsidies transparent. The framework also involves the clarification of property rights, the allocation of water to the environment, the adoption of trading arrangements in water, institutional reform and public consultation and participation.[104]

The separation of water property rights from land title, the adoption of trading arrangements, and public consultation and participation are significant in relation to Indigenous water rights. This will become more apparent in the following chapter.

The implementation of the reforms under the 1994 Framework was then tied to the National Competition Policy ('NCP').[105] Tying the reforms to the NCP 'gave the states and the Territories a financial incentive to meet these national policy goals'.[106] Thus various states amended their water legislation to achieve the Framework's objectives. Victoria, however, was one step ahead, having already enacted the *Water Act 1989* which included such Framework objectives as water trading[107] and allocation of water to the environment.[108]

A further two reports produced in the mid-1990s also influenced the policy direction of water reform. The first was a 1995 report called *Water Allocations and Entitlements: A National Framework for the Implementation of Property Rights in Water.* [109] Among other things, this report recognised 'ecological needs and the involvement of the community in planning processes'.[110] The other report was produced a year later and was titled *National Principles for the Provision of*

104 Council of Australian Governments, *Communiqué* (25 February 1994) <http://archive.coag.gov.au/coag_meeting_outcomes/1994-02-25/index.cfm>.

105 The National Competition Policy was the outcome of the Hilmer Report: Frederick G Hilmer, Mark R Rayner and Geoffery Q Taperell, *National Competition Policy* (Report, Independent Committee of Inquiry, 25 August 1993). This policy was then endorsed by COAG at its April 1995 meeting: Council of Australian Governments, *Communiqué* (11 April 1995) Attachment A – Agreement to Implement the National Competition Policy and Related Reforms <http://archive.coag.gov.au/coag_meeting_outcomes/1995-04-11/docs/attachment_a.cfm>.

106 Kildea and Williams, 'The Constitution and the Management of Water in Australia's Rivers', above n 67, 599.

107 Poh-Ling Tan, 'Irrigators Come First: Conversion of Existing Allocations to Bulk Entitlements in the Goulburn and Murray Catchments, Victoria' (2001) 18(2) *Environmental and Planning Law Journal* 154, 165–6.

108 Ibid 174, 180; Anita Foerster, 'Victoria's New Environmental Water Reserve: What's in a Name?' (2007) 11(2) *Australasian Journal of Natural Resources Law and Policy* 145, 155.

109 Agricultural and Resource Management Council of Australia and New Zealand, *Water Allocations and Entitlements: A National Framework for the Implementation of Property Rights in Water* (Occasional Paper No 1, Task Force on COAG Water Reform, October 1995).

110 Tan et al, *Collaborative Water Planning*, above n 72, 28.

Water for Ecosystems.[111] The principles contained in this report were endorsed by all Australian governments,[112] and generally speaking, 'gave pre-eminence to ecological values of water'[113] in order to resolve tensions between consumptive and non-consumptive uses. Of note is principle 12, which states that '[a]ll relevant environmental, social and economic stakeholders will be involved in water allocation planning and decision-making on environmental water provisions'. Neither report referred specifically to Indigenous involvement, but they expanded the concept of public consultation and participation, which had appeared in the 1994 Framework, to encompass community and stakeholder participation in decision-making about the management of water resources.

Not surprisingly, given its history, the MDB was still a significant focus of attention during this time. In 1995, following an audit of water use in the MDB that 'confirmed increasing levels of diversions and the consequent decline in river health',[114] a temporary cap on extractions was introduced, made permanent in 1997. The main goal of the cap 'was to establish a limit to extractions so as to protect the environment'. The implementation of the cap was an important decision[115] as without it there would have been 'a significantly increased risk that the environmental degradation of the river system of the Murray-Darling Basin would have been worse'.[116] The MDB is the main subject of the *Water Act 2007*, and is arguably one of the first real focal points for Indigenous participation in water management.

Intergovernmental Agreement on a National Water Initiative

Ten years after the 1994 Framework, it was recognised that despite its good intentions, it 'did not, as a whole, represent a coherent framework for water management'.[117] Thus, the states, territories and the Commonwealth reinvigorated the reform agenda by entering into the *NWI*. Described as 'represent[ing] the most significant innovation in water policy in more than a century',[118]

111 Agricultural and Resource Management Council of Australia and New Zealand and Australian and New Zealand Environment and Conservation Council, *National Principles for the Provision of Water for Ecosystems* (Occasional Paper SWR No 3, Sustainable Water Resources Management Committee Subcommittee on Water Resources, July 1996).
112 Tan et al, *Collaborative Water Planning*, above n 72, 28.
113 Ibid.
114 Murray–Darling Basin Commission, *Review of the Operation of the Cap* (Overview Report, August 2000) 7.
115 D J Blackmore, 'Managing for Scarcity: Water Resources in the Murray Darling Basin' (2000) *CEDA Bulletin* 50, 50.
116 Daniel Connell and R Quentin Grafton, 'Water Reform in the Murray-Darling Basin' (2011) 47(12) *Water Resources Research* 1, 3; Murray–Darling Basin Commission, above n 114, 14.
117 National Water Commission, *NWI Assessment 2011*, above n 98, 4.
118 Daniel Connell, Stephen Dovers and R Quentin Grafton, 'A Critical Analysis of the National Water Initiative' (2005) 10(1) *Australasian Journal of Natural Resources Law and Policy* 81, 85.

implementation of the objectives of the *NWI* is intended to 'result in a nationally-compatible, market, regulatory and planning based system of managing surface and groundwater resources for rural and urban use that optimises economic, social and environmental outcomes'. Unlike the 1994 Framework, the *NWI* is not tied to competition policy.[119]

Monitoring of the implementation of the *NWI* was undertaken by the National Water Commission ('NWC'), set up by the Commonwealth pursuant to the *National Water Commission Act 2004* (Cth).[120] The NWC produced four assessments of the implementation of the *NWI*, most recently in 2014. Despite its limitations, such as a lack of enforceability,[121] the aspirational nature of its terms,[122] and difficulties identified with its implementation,[123] the *NWI* remains the leading policy document on water management in Australia.

The most significant feature of the *NWI* in the context of this book is that it was the first national policy document following *Mabo* and the *Native Title Act 1993* (Cth) to explicitly recognise Indigenous water rights,[124] the 1994 Framework having remained silent on this issue.

The Water Act 2007 (Cth)

The *Water Act 2007* has been described as 'the most extensive Commonwealth intervention into water resource management in Australia since Federation'.[125] Malcolm Turnbull, then Minister for the Environment and Water Resources, stated: 'This Water Bill is the first water reform program introduced into this parliament in 106 years. It is truly a nation-building bill, not only for this generation but also for the generations to come.'[126]

The *Water Act 2007* arose out of then Prime Minister Howard's National Plan for Water Security.[127] The National Plan aimed to 'improve water use efficiency and address overallocation of water'.[128] Its focus was largely on, although not limited to, the MDB. It was intended that the Basin states would reach agreement on the National Plan and refer power to the Commonwealth to enact the *Water*

119 Gardner, Bartlett and Gray, above n 2, 53.
120 The NWC was abolished in 2015. Monitoring of the *NWI* is now done by the Productivity Commission.
121 Gardner, Bartlett and Gray, above n 2, 52.
122 For example, cl 52(i) regarding the inclusion of Indigenous representation in water planning contains the qualifier 'whenever possible'.
123 NWC, *NWI Assessment 2011*, above n 98, 4–5.
124 See, eg, *NWI*, cls 52–4. See Chapter 3 for more detail on the *NWI*.
125 Paul Kildea and George Williams, 'The Water Act and the Murray-Darling Basin Plan' (2011) 22 *Public Law Review* 9, 9.
126 Commonwealth, *Parliamentary Debates*, House of Representatives, 8 August 2007, 7 (Malcolm Turnbull, Minster for the Environment and Water Resources).
127 John Howard, *A National Plan for Water Security: 25 January 2007* (Commonwealth of Australia, 2007) <http://pandora.nla.gov.au/pan/10052/20070321-0000/www.pm.gov.au/docs/national_plan_water_security.pdf>.
128 Gardner, Bartlett and Gray, above n 2, 54.

Act 2007 under s 51(xxxvii) of the *Constitution*. [129] However, Victoria did not agree. Accordingly, the Commonwealth had to rely on other powers to enact the 2007 legislation – an apparent retreat from cooperative federalism.

This retreat was not to last long. The *Water Act 2007* was subsequently amended in 2008 after all the Basin states, including Victoria, signed the Murray–Darling Reform Agreement and legislated to refer their powers to the Commonwealth.[130] Pursuant to the 2008 amendments, a new Murray–Darling Basin Agreement was reached.[131]

The *Water Act 2007* is important in that, like the *NWI*, it too made some attempt to recognise Indigenous water rights, discussed in more detail in the next chapter.[132]

Concluding remarks

There are some notable water management themes that emerge from the above discussion.

Common law water rights, existing in the early years of the colonies, relied on being in possession of land. As Indigenous people were not recognised as being in possession of land, they were excluded from the common law water rights regimes. Emerging in the late nineteenth and early twentieth centuries were statutory innovations which vested the use, flow and control of water in the state and converted common law rights into statutory rights. Regulatory mechanisms were designed to facilitate irrigation and agriculture, not Indigenous uses. Then came the innovations of the 1960s and 1970s which began to treat water as a unified resource, and to recognise environmental concerns. The 1990s, however, were characterised by reforms of a micro-economic nature. This was also the era that coincided with the native title era.

The important observation to be made about these approaches to water management is that they all ignored Indigenous interests. It is only in the new millennium that this has started to change. The following chapter will therefore investigate how the regulatory regimes were developed, to ascertain why Indigenous interests were ignored, and why this has now started to change.

129 For an explanation of this mechanism, see Carney, above n 87, 14–17.
130 Gardner, Bartlett and Gray, above n 2, 55.
131 *Water Act 2007* sch 1.
132 Ibid ss 21(4)(c)(v), 22(1), 204(3)(a).

3 Indigenous participation in the development of Australian water management regimes

In the early days of the colonies, Indigenous people were not involved in the development of legislation, and therefore there is no mention in the early water legislation of Indigenous rights or interests.

Much has already been written about Australia's Indigenous people following colonisation,[1] their relationship with government and the laws that governed them, so it is not proposed to go into those matters in any detail. The following discussion highlights how some of those laws and relationships impacted on Indigenous people's ability to influence the development of Australia's water management regime, with the focus on Victoria, being our state-based case study. Much of it should come as no surprise to anyone with an understanding of Indigenous issues, but it is worth setting out as it provides some background and context to the modern era.

Indigenous people, as the initial inhabitants of Australia, occupied and possessed the country pursuant to their traditional laws and customs.[2] But, as noted earlier, Australia inherited English common law upon colonisation, and it was this law that governed concepts relating to land and waters rather than Indigenous concepts. This was a denial of Indigenous sovereignty, perpetuated through the common law doctrine of 'desert and uncultivated' relating to colonised land (or *terra nullius* at international law); a denial of sovereignty which even *Mabo* did not displace.[3] This denial of Indigenous sovereignty, and the associated treatment of Indigenous people generally as aliens in their own land and as lesser human beings than the colonisers, underpins all the factors outlined below as affecting

1 Richard Broome, *Aboriginal Victorians: A History Since 1800* (Allen & Unwin, 2005); Bain Attwood, *The Making of the Aborigines* (Allen & Unwin, 1989); Henry Reynolds, *The Other Side of the Frontier: Aboriginal Resistance to the European Invasion of Australia* (University of NSW Press, first published 1981, 2006 edn); Henry Reynolds, *Frontier: Aborigines, Settlers and Land* (Allen & Unwin, first published 1987, 1996 edn); Henry Reynolds, *With the White People* (Penguin, 1990); Richard Broome, *Aboriginal Australians: A History Since 1788* (Allen & Unwin, 4th edn, 2010); Heather McRae et al (eds), *Indigenous Legal Issues, Commentary and Materials* (Lawbook, 4th edn, 2009) ch 1.

2 *Mabo* (1992) 175 CLR 1.

3 Bruce Buchan, '"Aboriginal Welfare" and the Denial of Indigenous Sovereignty' (2002) 20 *Arena Journal* 97, 98.

Indigenous people's ability to influence the colonial government's response to water management issues.

How, then, did one influence the development of legislation generally, and, in our case, water management legislation? In the early days of the colonies, the main sources of influence were the power of the ballot box, lobbying strength (including the importance of the issue being lobbied) and, more directly, being a member of parliament.[4]

Voting rights

Victoria separated from New South Wales in 1851,[5] but it was not until the enactment of the *Constitution Act 1855* [6] that Victorians obtained their own parliament and became entitled to vote as a separate colony. Possession of an interest in land, whether freehold, leasehold or licence, was an important factor in a person's ability to stand for office and have an influence on the government of the day. So Indigenous people were at an immediate disadvantage, having been dispossessed of their land.

However, there were other prerequisites to satisfy. To be able to vote, one was required to be able to read and write. There was also a residency requirement, although if one was resident at a charitable institution, such as a mission station, this itself was an impediment.[7] Women were not able to vote. Prisoners and those who had not paid their rates and taxes were disqualified from voting. Aboriginal men were not directly disqualified from either standing for office or voting, but these prerequisites, along with various other factors, all affected their ability to be able to exercise these rights.[8]

The power of the Aboriginal vote

But even if every Aboriginal male over the age of 21 was literate, had a place of residence that was not on a mission station, and was registered to vote, the sheer numbers of non-Aboriginal voters would mean that the Aboriginal vote would be

4 Religious groups and the media also influenced the government of the day: Raymond Wright, *A People's Counsel – A History of the Parliament of Victoria: 1856–1990* (Oxford University Press, 1992) 69–71.

5 *An Act for the better Government of Her Majesty's Australian Colonies 1850* (Imp) better known as the *Australian Constitutions Act 1850* (Imp). Pursuant to writs issued by New South Wales under this Act, Victoria became a separate colony on 1 July 1851.

6 The *Constitution Act 1855* was enacted as a schedule to a British Act of Parliament: *An Act to enable Her Majesty to assent to a Bill, as amended, of the Legislature of Victoria, to establish a Constitution in and for the Colony of Victoria* [18 & 19 Vic. c.55].

7 *Electoral Act 1863* (Vic) s XII.

8 For further detail on Aboriginal people and the vote, see, eg, Murray Goot, 'The Aboriginal Franchise and its Consequences' (2006) 52(4) *Australian Journal of Politics and History* 517; John Chesterman and Brian Galligan, *Citizens Without Rights: Aborigines and Australian Citizenship* (Cambridge University Press, 1997) ch 1.

a minority vote of negligible influence. Add to this the fact that voting was not compulsory in Victoria until 1926,[9] so Aboriginal people would also need to have been convinced that there would be some benefit to them arising from their participation in the political processes of the colonisers.

Just to highlight how futile the Aboriginal vote would have been had it been exercised, population data from 1891 indicate that there were only 325 Aboriginal males counted out of a total male population of 598,414.[10] Importantly, these figures are from the period of significant legislative developments in water management in Victoria, developments that were to shape the continuing management of water resources for over a century.

That is not to say that Aboriginal people were ignored. On the contrary, historian Bain Attwood notes that 'Victoria was the colony where Aborigines were given the most sustained political attention in the nineteenth century'.[11] Indeed, there was so much concern for the declining Aboriginal population of Victoria and their treatment by the settlers that in 1838 the Port Philip Protectorate was established.[12] The Protectorate was abolished in 1849, although the concern continued, and in 1860 a 'Central Board to watch over the interests of the Aborigines in the colony of Victoria' was established,[13] followed in 1869 by the enactment of the *Aborigines Protection Act.* [14]

However, by 1886, when the government enacted the notorious *Aborigines Protection Act 1886* (Vic) ('*APA 1886*') Aboriginal people had become much less of a political priority,[15] having been overshadowed by more important considerations such as irrigation and water supply. Unlike the Water Supply and Irrigation Bill 1886 (Vic), about which there was a fierce and lengthy debate[16] and which ultimately became the *Irrigation Act 1886* (Vic), the Aborigines Protection Law Amendment Bill of 1886 received little and only last-minute attention in parliament.[17] As one member of the Legislative Assembly observed during the very limited debate, 'he had heard before of hasty legislation', but this was 'hasty legislation with a vengeance'.[18] The following day, *The Argus* reported the debate as follows:

9 This was in relation to the Legislative Assembly under the *Compulsory Voting (Assembly Elections) Act 1926* (Vic). In the Legislative Council, voting did not become compulsory until 1935 by virtue of the *Legislative Council Elections Act 1935* (Vic).

10 Government Statist, *Census of Victoria, 1891: Parts I to IX, with a General Report by the Government Statist, Taken on the 8th April, 1891* (1893) 22, 25.

11 Attwood, *The Making of the Aborigines*, above n 1, 81.

12 See, eg, Jane Miller, 'A Guide to Government Acts, Reports and Regulations Relating to Indigenous Victorians' (2010) 85 *La Trobe Journal* 172, 173.

13 Ibid 176.

14 Victoria was the first of the colonies to enact such legislation: Buchan, above n 3, 116; Attwood, *The Making of the Aborigines*, above n 1, 84.

15 Attwood, *The Making of the Aborigines*, above n 1, 97.

16 David Ingle Smith, *Water in Australia: Resources and Management* (Oxford University Press, 1998) 153.

17 Attwood, *The Making of the Aborigines*, above n 1, 81.

18 Victoria, *Parliamentary Debates*, Legislative Assembly, 15 December 1886, 2913 (J T Brown).

> Members of the Legislative Assembly spent the last night of the session in a hilarious mood, as usual. The subjects before them changed so rapidly that fixed thought on any topic was out of the question, and time passed quite pleasantly in consequence. … In the last 10 minutes left to the Government Mr Deakin scored by pushing the Aborigines Protection Bill through all its stages. Some members were irate, and protested against hasty legislation, just as some members protest on the last night of every session, but their voices were hardly heard in the din.[19]

An editorial in *The Age*, two days after the passage of the Aborigines Protection Law Amendment Bill, is indicative of the importance of the Water Supply and Irrigation Bill vis-à-vis the other Bills passed in the parliamentary session:

> The other Bills which have been carried through their final stages are not merely of a minor character when compared with those relative to water supply and irrigation, but would not have stood out as important in any session.[20]

Although considered to be of 'minor character' which 'would not have stood out as important' to non-Indigenous Victorians (if the editorial in *The Age* is any indication of this), the Bill which resulted in the *APA 1886* was to become of major significance to Aboriginal Victorians in the years to follow, being arguably 'the most draconian Aboriginal legislation of its time in Australia'.[21]

The *APA 1886* has been described by Broome as 'a deliberate policy of absorption' whereby 'the Aboriginal race would vanish as the "full bloods" died out and the "half-castes" were blended to whiteness'.[22] The view that Aboriginal people were a dying race was based on the theory of social Darwinism which had been gaining popularity in the latter half of the nineteenth century.[23] In 1881 *The Age* reported:

> It appears to be taken for granted that the aborigines are doomed. They must, like the aborigines of other countries, pass away before the white man. It would be useless at this period to discuss the cause of this or deny the inevitable.[24]

19 Editorial, *The Argus* (Melbourne), 16 December 1886, 7.
20 Editorial, *The Age* (Melbourne), 17 December 1886, 4.
21 Broome, *Aboriginal Victorians*, above n 1, 181. For more detail on why this legislation 'was the most draconian … of its time', see ch 10. See also Attwood, *The Making of the Aborigines*, above n 1, ch 4.
22 Broome, *Aboriginal Australians*, above n 1, 94.
23 '[T]he theory that societies, classes, and races are subject to and a product of Darwinian laws of natural selection. Often used to justify political conservatism, imperialism, and racism.' *Oxford English Dictionary Online* (Oxford University Press, 2018) <www.oed.com>.
24 Editorial, *The Age* (Melbourne), 13 January 1881, 2.

It made a similar observation around seven years later.[25]

Thus, if the Aboriginal race was perceived as dying out, then there would be clearly no need to listen to Aboriginal views about their long-term future and, more specifically here, the continuing (long-term) management of water resources.

Lobby groups

That is not to say that Aboriginal people remained silent and quietly disappeared as predicted. There are notable instances during this time of Aboriginal Victorians lobbying for land and better conditions, the Coranderrk protests being a prime example.[26] Attwood has noted that '[t]hey [the Coranderrk Aboriginal residents] were adept students of the colonial political system and adroit exponents of its techniques of lobbying'.[27] This lobbying had some success; following an enquiry into the condition and management of the station in 1881, motivated to a large extent by the lobbying efforts,[28] Coranderrk was permanently reserved in 1884.[29] The success was to be short-lived, though; the permanent reservation did not protect the land from later encroachment by non-Indigenous interests and was eventually sold in 1948.[30]

Edwyna Harris has suggested that lobbying was a factor of particular relevance to the development of water management institutions in Victoria between 1850 and 1886.[31] She contends that '[t]he effectiveness of a group's lobbying effort is positively related to its size and wealth. In this framework, wealth does not simply mean financial resources but can relate to other resources at their disposal such as votes.'[32] It is also 'inversely related to the size and wealth of opposing groups'.[33] Harris then argues that lobbyists in Victoria were particularly effective in

25 Editorial, *The Age* (Melbourne), 11 January 1888, 4.

26 Diane Barwick, *Rebellion at Coranderrk* (Aboriginal History, 1998); See also Broome, *Aboriginal Victorians*, above n 1, ch 9; Bain Attwood, *Rights for Aborigines* (Allen & Unwin, 2003) ch 1.

27 Attwood, *Rights for Aborigines*, above n 26, 19. See also Broome, *Aboriginal Victorians*, above n 1, 169.

28 Attwood, *Rights for Aborigines*, above n 26, 21.

29 Broome, *Aboriginal Victorians*, above n 1, 180.

30 This occurred by virtue of the *Coranderrk Lands Act 1948* (Vic). In 1991, the Coranderrk mission cemetery was granted to the Wurundjeri Tribe Land and Compensation Cultural Heritage Council pursuant to the *Aboriginal Lands Act 1991* (Vic). In 1998 a small portion of the original station was purchased by descendants of the Corranderrk community, and is held in title by the Wandoon Estate Aboriginal Corporation: Indigenous Land Corporation, *Land Purchased* <www.ilc.gov.au/Home/What-We-Do/Land-Purchased>.

31 Edwyna Harris, 'Institutional Change and Economic Growth: The Evolution of Water Rights in Victoria, Australia 1850–1886' (2007) 26(2) *Economic Papers* 118.

32 Ibid 120.

33 Ibid.

> bringing the Government's attention to the need for institutional change to support irrigation investment and agricultural expansion. Political consensus regarding the need for change was reached relatively quickly because lobbying efforts took place in the absence of any other critical vested interest groups whose demands had to be met.[34]

These lobbyists would have consisted largely of irrigators and farmers. Following the end of the mining boom in the mid-1860s, the government was intent on economic development to provide employment for thousands of miners looking for work, and agricultural expansion in the guise of the 'yeoman farmer'[35] was a means by which this could be achieved.[36]

These yeoman farmers made up a large sector of the voting population and had similar interests, enabling them to join forces in putting their views to the government. Although the urban population increased at a greater rate than the rural population in the second half of the nineteenth century, the rural population remained a significant voting force.[37] And with voting power came lobbying influence.

Although Harris does not consider Indigenous interests, Indigenous Victorians would clearly be an interest group by today's standards. But, as noted above, Indigenous people had low levels of literacy, were a tiny minority of the voting-age population (if indeed they voted at all), and few, if any, would have possessed land according to the common law tradition inherited from England. In the government's eyes, therefore, they were not, to borrow Harris's phrase, a 'critical vested interest group whose demands had to be met'. Rather, they were considered a doomed race of people who were 'incapable of leading an autonomous life and becoming contributing members of the society'.[38]

One could surmise that perhaps Indigenous interests in water might have been taken notice of had they been similar to non-Indigenous interests, such as the use of water for irrigation or farming. It has been argued that this occurred in NSW in

34 Ibid 122.

35 Yeoman – 'a countryman, especially one of some social standing, who cultivates his own land'. *Macquarie Dictionary Online* (Macquarie Dictionary Publishers, 2018) <www.macquariedictionary.com.au>.

36 Sandford D Clark and Ian A Renard, *The Law of Allocation of Water for Private Use, Volume 1: The Framework of Australian Water Legislation and Private Rights* (Research Project 69/16, Australian Water Resources Council, 1972) 156. See also Harris, above n 31, 121.

37 This was due to the deliberate practice of rural malapportionment whereby 'seats in rural and/or regional zones would be entitled to a lower enrolment than seats in urban and metropolitan zones': Nick Economou and Zhareh Ghazarian, *Electoral Redistributions in Theory and Practice* (Report, Victorian Electoral Commission/ Monash University, November 2006) 15. For more information on the basis for this practice, see Economou and Ghazarian, above n 37, 11.

38 Miller, above n 12, 172.

relation to Aboriginal demands for land.[39] But circumstances in Victoria suggest that even farming using European methods would not necessarily have guaranteed Aboriginal people any influence on government.

Coranderrk, for example, was established in light of a commitment by Aboriginal people to farm it using 'white man' techniques,[40] and for a while was relatively successful as a farm. However, due to circumstances largely beyond the control of the residents, it began to decline,[41] and in the early 1890s was reduced in size by almost half.[42] This reduction in the size of the station occurred despite it having been made a permanent reserve in 1884.[43]

Thus, despite efforts to farm the station like Europeans, and some significant lobbying efforts of both the Coranderrk Aboriginal residents and their non-Indigenous supporters,[44] it appears there were too many other factors working against the Coranderrk residents for them to have any continuing influence with government. Those factors included lack of secure tenure, the opposing interests of settlers wanting to farm the land themselves,[45] and a belief, based on the theory of social Darwinism, that Aboriginal people were a dying race.

This discussion has largely been in the context of land, but the points being made are equally applicable to water, given the nexus between land and water in the physical sense, as well as legally, both in the common law and the legislation evolving at the time. In any event, Indigenous interests in both land and water involve a much wider spectrum of interests than the largely economic interests of the early pastoralists and squatters. Land and water had (and still have) a spiritual and cultural dimension of which some of the settlers may have been aware, but would almost certainly not have understood.

In summary, the ability for Aboriginal people to influence the development of Victoria's water management regime was potentially via voting and lobbying. Impediments to their ability to vote were a lack of literacy, residency requirements, not being in possession of land, and being in receipt of charity. Even if those impediments were overcome, Aboriginal people were such a small proportion of the population that their vote would have had no impact. Lobbying only got them so far, but as they were considered to be in need of protection and a 'dying race', their views were treated as if of negligible importance.

39 Heather Goodall, 'Land in Our Own Country: The Aboriginal Land Rights Movement in South-Eastern Australia, 1860–1914' (1990) 14(1) *Aboriginal History* 1, 10.

40 Letter from William Thomas, Guardian of Aborigines, to Brough Smyth, R, Secretary of the Central Board appointed to watch over the interests of the Aborigines of the Colony of Victoria, 26 July 1860. Quoted in Aldo Massola, *Coranderrk: A History of the Aboriginal Station* (Lowden Publishers, 1975) 8.

41 For more information on the decline of Coranderrk, see Broom, *Aboriginal Victorians*, above n 1, ch 9.

42 Massola, above n 40, 35.

43 As a permanent reserve it could only be changed by an Act of Parliament. This shows that even tenure guaranteed by statute does not guarantee security of tenure.

44 Attwood, *Rights for Aborigines*, above n 26, 20.

45 Broome, *Aboriginal Victorians*, above n 1, 169.

Indigenous participation in the development of water management regimes post-federation

The previous discussion concentrated on Indigenous influence (or lack thereof) in the nineteenth century. Let us now consider the next 100 years or so. The focus here will be oriented more specifically towards Indigenous influence on the development of water management regimes, commencing with a review of Victoria's water management regime.

Victoria

Indigenous participation in the development of the Groundwater Act 1969 *(Vic)*

Following the enactment of the *Water Act 1905*, the main features of which were outlined earlier, water law reform in Victoria was incremental throughout most of the twentieth century, the only significant reform prior to the *Water Act 1989* being the enactment of the *Groundwater Act 1969*. Given the success of the 1967 constitutional referendum[46] and a clear momentum now on land rights around Australia bringing Indigenous issues to the forefront of the national consciousness,[47] the question arises as to whether Aboriginal Victorians, as potential landholders pursuant to emerging land rights, played any role in the development of the *Groundwater Act*. The short answer is no. However, Aboriginal people were not alone in that regard; there is no evidence of any public or stakeholder participation, let alone Aboriginal participation, in the development of the *Groundwater Act*. Instead, the Act appears to have been based largely on recommendations arising out of a 1962 report into underground water resources prepared by the State Development Committee.[48]

It is now common, at both the Commonwealth and state level, to actively seek public and stakeholder input into the development of environmental policy and legislation.[49] But in 1962, the concept of public participation had not yet become accepted as a legitimate input. This concept did not take hold until the

46 This referendum amended the *Constitution* to enable the Commonwealth Parliament to make laws for Aboriginal people, and for Aboriginal people to be counted in the national population.

47 For example, South Australia passed the first land rights legislation in 1966, the *Aboriginal Lands Trust Act 1966* (SA); during the 1950s and 1960s Aboriginal people in Victoria were lobbying for the transfer of title to Lake Tyers, which they eventually obtained pursuant to the *Aboriginal Lands Act 1970* (Vic).

48 State Rivers and Water Supply Commission, *Underground Water Resources of Victoria: Report of the State Development Committee – Analysis of Recommendations* (Melbourne, 1962). See also Victoria, *Parliamentary Debates*, Legislative Assembly, 23 October 1968, 1235 (W A Borthwick, Minister of Water Supply).

49 Poh-Ling Tan, 'Legislating for Adequate Public Participation in Allocating Water in Australia' (2006) 31(1) *Water International* 12, 13.

1980–90s,[50] and is manifested in the international sphere in documents such as the *Rio Declaration.* [51]

Indigenous participation in the development of the Water Act 1989 *(Vic)*

The next major development in water resource management in Victoria was the *Water Act 1989.* This Act was the result of a comprehensive review of Victoria's water laws over a number of years – a review that was long overdue. Prior to its enactment there were 21 existing statutes relating to water, amounting to around 1,100 pages,[52] described in 1983 by the Minister for Water Resources as being 'a maze of antiquated, imprecise, overly complicated and often inconsistent legislation'.[53] As noted by the Department of Water Resources in 1986, '[w]ater legislation in Victoria has not had a comprehensive overhaul since 1905, so it is not surprising that its provisions do not reflect contemporary water management needs'.[54] The draft Water Bill arising out of the review proposed to consolidate and replace 15 of those Acts with a single statute.[55]

The *Water Act 1989* was developed with major input from stakeholders and the public. It commenced with an internal departmental review team which 'analysed the accumulated material and canvassed what would be required of a modern legislative framework'.[56] The results were then published in September 1986 as a Discussion Paper, being 'the first public document of the review process'.[57] Mulligan and Pigram describe the ensuing public review process as follows:

> The review was divided into three main sections: the legal framework for water sector management; resource use; and protection of individual and community interests. A series of Issues Papers dealing with all aspects of the Water Law Review were distributed widely throughout the State to interested individuals and key organisations. These were followed by a number of regional public meetings in February and March 1987 to discuss issues, provide explanations, and gain comments on the proposals. Over 500 written

50 Dingle Smith, 'Water Resources Management' in Stephen Dovers and Su Wild River (eds), *Managing Australia's Environment* (Federation Press, 2003) 53, 68.

51 *Report of the United Nations Conference on Environment and Development, Volume I, Resolutions Adopted by the Conference*, UN Doc A/CONF.151/Rev.1 (Vol 1) (adopted 14 June 1992) Annex I. Principle 10 relates to public participation and principle 22 relates to Indigenous participation.

52 Helen Mulligan and John J Pigram, *Water Administration in Australia: Agenda for Change* (Occasional Paper No 4, University of New England, 1989) 42.

53 Victoria, *Parliamentary Debates*, Legislative Council, 2 December 1983, 1566 (David White, Minister for Water Resources).

54 Department of Water Resources Victoria, *Water Law Review* (Discussion Paper, Report No 1, Water Management Report Series, September 1986) 9–10.

55 Independent Committee of Review, *Water Law Review: Draft Proposals for a Water Bill* (Report, Department of Water Resources Victoria, January 1989) 1, 3.

56 Ibid 1.

57 Ibid.

> submissions, as well as the comments raised in the public meetings, were analysed and taken into account in drafting the new legislation.[58]

A report on the outcomes of the review process suggests that there were no Indigenous people or organisations which put in a submission.[59] The fact that the *Water Act 1989* as enacted made no reference to Indigenous interests in water supports a conclusion that Indigenous people did not participate in the public review process.

This should not be surprising given that the acknowledgement of Indigenous rights and interests as requiring separate consideration in matters relating to natural resource management had yet to take hold in Australia. This was not to occur until the early 1990s when it started appearing in various national policy documents (about which more is said below), this being also the period in which *Mabo* [60] was decided and the *NTA* enacted in response. But what is more surprising is that despite these policy and legal developments, a review of Victoria's water management regime in 2003–04 did not result in any amendments to the *Water Act 1989* to recognise Indigenous interests in water. So what was involved in this later review and why weren't Indigenous interests recognised?

In August 2003, the Victorian Labor Government released a *Green Paper* on water reform.[61] The purpose of the *Green Paper* was to promote community discussion on the best way to manage Victoria's water resources in the face of increased pressures on water supply, which had been 'brought sharply into focus by the current severe drought'.[62] The *Green Paper* made only one reference to Indigenous people,[63] but none of the principles or proposals put forward for discussion in the *Green Paper* mentioned Indigenous interests.

The *Green Paper* resulted in the Victorian government's water policy, *Victorian Government White Paper – Securing Our Water Future Together* ('*White Paper*'), released in June 2004.

A media release on 16 January 2004 noted that 'over 10,000 copies of the *Green Paper* have been distributed and around 2,500 people have attended 19 open public information forums held across Victoria during October and November'.[64] Details of who put in submissions are now scant, but given the wide-ranging consultation process, it is possible that Indigenous people made submissions on the *Green Paper* or attended one of the public meetings. If so, it would appear, however, that the Indigenous voice was drowned out by the public response, there being only three limited references to Indigenous people in the

58 Mulligan and Pigram, above n 52, 42 (citations omitted).
59 Independent Committee of Review, above n 55, Appendix 4.
60 (1992) 175 CLR 1.
61 Victorian Government, *Securing Our Water Future: Green Paper for Discussion* (Department of Sustainability and Environment, August 2003).
62 Ibid 5.
63 Ibid 13.
64 Office of the Premier (Vic), '*Green Paper* Submissions Now Available on the Web' (Media Release, 16 January 2004).

White Paper, focussing on non-legislative recognition and 'dialogue'.[65] Accordingly, amendments to the *Water Act* resulting from the *White Paper* did not include any acknowledgement of Indigenous interests.

There are a couple of possible explanations for this lack of acknowledgement. A number of native title claims had been lodged in Victoria under the *NTA*, the most notable being the Yorta Yorta claim. After failing at first instance,[66] it went all the way on appeal to the High Court where it ultimately failed.[67] That decision was handed down in August 2002, a year before the *Green Paper* was released. It was a widely held view at the time (subsequently proven wrong) that the consequence of this decision would be that limited (if any) native title would be ever found in Victoria. And even though the Victorian government reached an in-principle agreement with the Wimmera Clans for a consent determination of native title in October 2002,[68] this agreement (and subsequent determination in 2005)[69] did not include any native title rights and interests in water – despite the determination area being essentially the beds and banks of various water bodies. This suggests that because of these two determinations, the government was assuming that no other Aboriginal group would have their native title rights and interests in water recognised under the *NTA*, and that only legally recognised water rights could be accommodated in the *Water Act 1989*.

Another factor in the lack of recognition of Indigenous interests through the 2003–04 review process is the size of the Indigenous voice. In Victoria at that time, Indigenous people were approximately 0.6 per cent of the population.[70] In addition, as a significantly disadvantaged sector of the community, the pressures of day-to-day living would have left little time or energy to devote to community consultation processes. So, without specific efforts by government to engage with Indigenous communities, there would have been little incentive for Indigenous people to participate.

Finally, Victoria (along with the rest of south-east Australia) was in the middle of a severe drought. This, as noted earlier, was the impetus behind the proposed reform of Victoria's water laws. Given that one of the objects of the *Water Act 1989* is 'to continue in existence and to protect all public and private rights to water existing before the commencement of the relevant provisions of this Act',[71] and with fierce competition for scarce water resources among those with existing rights, it was always going to be an uphill battle for Indigenous Victorians to have their rights recognised, particularly in light of the factors just outlined.

65 *White Paper* 8, 10, 26.

66 *Yorta Yorta v Victoria* [1998] FCA 1606 (18 December 1998).

67 *Yorta Yorta v Victoria* (2002) 214 CLR 422.

68 National Native Title Tribunal, *Wotjobaluk, Jaadwa, Jadawadjali, Wergaia and Jupagulk Native Title Determinations – What They Mean for the Wimmera Region* (December 2005) 3.

69 *Clarke v Victoria* [2005] FCA 1795 (13 December 2005).

70 Australian Bureau of Statistics, *4705.0 Population Distribution, Aboriginal and Torres Strait Islander Australians 2006* (2007) 5.

71 *Water Act 1989* s 1(m).

The very limited acknowledgement of Indigenous interests in the 2003–04 water reform papers, resulting in no amendments to the *Water Act 1989* in 2005, is also surprising given the amendments to Victoria's *Constitution Act 1975* (Vic) in 2004 which recognise Victoria's Aboriginal people as having 'a spiritual, social, cultural and economic relationship with their traditional lands *and waters* within Victoria'.[72]

However, in 2010, as a consequence of the enactment of the *Traditional Owner Settlement Act 2010* (Vic) ('*TOSA*'), the *Water Act 1989* was amended to include s 8A, which recognises rights that have been recognised in a *TOSA* agreement. This was the first time that Indigenous water rights had been recognised in Victoria's water laws. The *TOSA* is discussed in more detail in a later chapter, but it suffices to say here that such rights are limited and do not include management rights.

Recent water law reform in Victoria

Victoria commenced another review of its water legislation in late 2012 under a conservative government, which had won office in November 2010. The review process involved the production of an exposure draft bill in 2013 about which there were extensive consultations with stakeholders and the public, including Traditional Owners. Numerous submissions were received as part of this process, many of which sought better recognition of Traditional Owner water rights in the draft bill.[73] However despite the consultation with Traditional Owners and the many submissions in support of greater recognition of their rights, only one amendment was made to the exposure draft of the bill, and a very minor one at that.[74] An equivalent of s 8A of the *Water Act 1989* was retained.

So despite consultation with Traditional Owners, it would appear that the government took the view that it did not need to do any more. Consultation in this instance amounted to little more than a 'tick-a-box' exercise.[75]

The bill was introduced into the Victorian Parliament on 24 June 2014; however, due to a change in government back to Labor in November 2014, the bill lapsed.[76]

In 2016 the new Victorian government produced a wide-ranging strategic plan for the management of water resources. Called *Water for Victoria: Water Plan*, it

72 *Constitution Act 1975* (Vic) s 1A(2)(b) (emphasis added).
73 There were 146 submissions, 20 of which sought greater recognition of Indigenous rights.
74 The amendment was to the definition of cultural values that made it clear that cultural values included Indigenous cultural values.
75 Note that under s 24HA of the *NTA* (discussed in Chapter 4), there is no requirement to notify or give native title holders the opportunity to comment on the making, amendment or repeal of legislation relating to the management or regulation of water.
76 For more detail about the 2012–14 water law review, see Katie O'Bryan, *From Aqua Nullius to Aqua Minimus? The Legal Recognition in Victoria of Indigenous Rights to Participate in the Management of Water Resources – Lessons from Aotearoa-New Zealand* (PhD Thesis, Monash University, 2015) 87–92.

was developed with the input of Traditional Owners,[77] and has a dedicated chapter on recognising and providing for Aboriginal values.[78] This came about because the government acknowledged that up until then 'Aboriginal people have had limited involvement in decision-making in Victoria's water planning and management framework, leading to a lack of protection of Aboriginal water values and a failure to make use of traditional ecological knowledge'.[79]

From the *Water Plan* emerged proposals for amending Victoria's water laws to specifically acknowledge the role of Indigenous knowledge in water management by providing for Indigenous representation on various advisory committees, and to ensure that various plans and strategies have regard to Aboriginal cultural values and uses of waterways. The resulting Water and Catchment Legislation Bill 2017 was introduced into the Legislative Assembly of the Victorian Parliament on 31 October 2017 and is currently before the Legislative Council. If passed, it will progress the recognition of the role of Indigenous people in Victoria's water laws.

Occurring in parallel with the preparation of the *Water Plan* was the development and subsequent enactment of the *Yarra River Protection (Wilip-gin Birrarung murron) Act 2017* (Vic). This Act also progresses Indigenous interests in water management by first establishing an advisory body, the Birrarung Council, which must have at least two Traditional Owner representatives on it. Second, it contains principles to guide the protection of the Yarra River which highlight Aboriginal cultural values, heritage and knowledge, and the importance of involving Traditional Owners in policy planning and decision-making. Finally, the Act includes Traditional Owner language in both its title and preamble,[80] a first for Victorian law.

Political will was clearly a significant factor in facilitating these changes to Victoria's water laws.

The modern era of water reform and Indigenous participation – the national dimension

The *National Strategy for Ecologically Sustainable Development* of 1992 ('*NSESD*') was the first major national natural resource management policy document to recognise the need to consult with Indigenous people.[81] It has also been described

77 Department of Environment, Land, Water and Planning, *Water for Victoria: Closing the Loop Feedback Report* (2016).

78 Victorian Government, *Water for Victoria: Water Plan* (Department of Environment, Land, Water and Planning, 2016) ch 6.

79 *Water for Victoria: Discussion Paper* (Department of Environment, Land, Water and Planning, 2016) 87.

80 Woi-wurrung, the language of the Wurundjeri people, the Traditional Owners of much of the catchment of the Yarra River.

81 *National Strategy for Ecologically Sustainable Development* (December 1992) 82–3. The *National Forest Policy Statement – A New Focus for Australia's Forests* (1992) also referred to Indigenous interests.

as the 'first truly important contemporary national policy pronouncement'[82] in relation to (among other things) water resources management.

Although the final *NSESD* recognises the need to consult with Indigenous people,[83] it was noted that there were problems in its development in terms of the treatment of Indigenous issues and Indigenous scope for input.[84] One can deduce from the various documents produced as part of the process that this was indeed the case.

The process involved a discussion paper (which was silent on Indigenous interests) and the establishment of various working groups, each of which produced a report to inform the *NSESD*. All but two of the draft reports were released for public comment, and workshops were held to discuss those two reports. There was no Indigenous representation on any of the working groups, and only limited consultation with Indigenous organisations. However, a background paper on Aboriginal involvement was circulated for discussion at the two workshops,[85] and Indigenous issues were raised in some of the final reports.[86]

Although the *NSESD* states that it evolved 'through extensive consultation',[87] there was some criticism of the consultation process, particularly in relation to the working groups.[88]

The final reports emanating from this process formed the basis of the draft *NSESD* that was put out for public comment.

The *NSESD* was endorsed by the Council of Australian Governments ('COAG') at its very first meeting on 7 December 1992.[89] It is ironic that the *NSESD*, although promoting Indigenous consultation, was itself developed with only minimal Indigenous consultation.[90]

82 Alex Gardner, Richard Bartlett and Janice Gray, *Water Resources Law* (LexisNexis Butterworths, 2nd edn, 2018) 46.

83 *NSESD*, above n 81, objective 22.2.

84 Stephen Dovers, 'Discrete, Consultative Policy Processes: Lessons from the National Conservation Strategy for Australia and National Strategy for Ecologically Sustainable Development' in Dovers and Wild Rivers (eds), above n 50, 142, 145.

85 Ecologically Sustainable Development Working Group Chairs, *Intersectoral Issues Report* (January 1992), 246, Appendix A: List of Background Papers Prepared for Intersectoral Workshops, 91/17, Ros Sultan, and Rosie Beaumont (ACF), *Aboriginal Involvement Discussion Paper*.

86 See Ecologically Sustainable Development Working Groups, *Final Report – Mining* (November 1991) 255, 256; Ecologically Sustainable Development Working Groups, *Final Report – Fishing* (November 1991); Ecologically Sustainable Development Working Groups, *Final Report – Tourism* (November 1991).

87 *NSESD*, above n 81, 12.

88 See, eg, Mark Diesendorf, 'Reclaiming the Ecologically Sustainable Development Process' (1992) 13(3) *Bogong* 20; Bill Hare, *Ecologically Sustainable Development: Assessment of the ESD Working Group Reports* (Australian Conservation Foundation, 1991) 7.

89 Council of Australian Governments, *Communiqué* (7 December 1992) <http://archive.coag.gov.au/coag_meeting_outcomes/1992-12-07/index.cfm#environment>.

90 For more detail on the development of the *NSESD*, see O'Bryan, above n 76, 92–4.

The 1994 COAG Agreement on a Water Reform Framework

Following in the footsteps of the *NSESD*, and after *Mabo* and the enactment of the *NTA*, came the 1994 COAG Agreement on a Water Reform Framework ('1994 Framework'). Despite these developments, the 1994 Framework did not mention Indigenous interests, so its silence in that regard is noteworthy.

The Framework's silence may be attributed to the fact that its reforms were, apart from those relating to the environment, essentially micro-economic reforms of the utility sector, as indicated by the nature of various reports that informed the Framework.[91] Those thought to be most affected by the reforms were those who had existing water allocations. Given that Indigenous people did not have water allocations, their interests were likely either to be beyond the purview of those who drafted the Framework, or thought to be accommodated by the environmental provisions.

Although apparently the 1994 Framework (along with various other policy statements) 'was preceded by long periods of gestation and consultation',[92] there is no evidence to suggest that Indigenous people were specifically consulted in the lead up to its adoption.

The National Water Initiative

As noted earlier, the *National Water Initiative* ('*NWI*') was the first national policy document to specifically recognise Indigenous interests in water. However, it has been suggested that there was no Indigenous input into the development of the *NWI*. [93] In that regard, there is very little documentation setting out the consultation processes involved in developing the *NWI*. A discussion paper was released on 19 March 2004 and a series of consultation meetings were held between 31 March and 7 April 2004 with key stakeholders in all state capitals and Canberra.[94] There is also some evidence of environmental groups putting in submissions to the discussion paper, some of which make reference to Indigenous interests.[95] The discussion paper itself does not appear to be in existence any

91 See, eg, Industry Commission, *Water Resources and Waste Water Disposal* (Report No 26, 17 July 1992); Frederick G Hilmer, Mark R Rayner and Geoffrey Q Taperell, *National Competition Policy: Report of the Independent Committee of Inquiry* (25 August 1993); Eric Neal, *Report of the Working Group on Water Resource Policy to the Council of Australian Governments* (1994).

92 Smith, *Water in Australia*, above n 16, 269.

93 Sue Jackson, *National Indigenous Water Planning Forum: Background Paper on Indigenous Participation in Water Planning and Access to Water. A Report Prepared for the National Water Commission* (CSIRO, 2009) 12. It also appears that consultation of any form was lacking; see Tim Fisher, 'Water Sustainability or Sell-out? The National Water Initiative in Perspective' (2004) 36 *Australian Options* 20, 25.

94 Indicator: IW-43 Implementation of National Water Initiative, Department of Sustainability, Environment, Water, Population and Communities (Cth) <http://webarchive.nla.gov.au/gov/20110328054205/http://environment.gov.au/soe/2006/publications/drs/indicator/212/index.html>.

95 Environment Victoria, *Submission in Response to the Discussion Paper, National Water Initiative*, April 2004; Nature Conservation Council of New South Wales, *Submission*

more, nor is there any information setting out who attended the key stakeholder meetings. However, this does not preclude the possibility that there was some Indigenous consultation.

For example, the outcomes of the COAG meeting on 29 August 2003, at which the states and the Commonwealth agreed to develop the *NWI*, state that 'Governments will also consult with stakeholders representing industry, environment, local government and ***Indigenous*** interests'.[96] On this basis, it could be assumed that Indigenous people were invited to participate in the stakeholder consultation meetings held between 31 March and 7 April. In that regard, Aboriginal and Torres Strait Islander Commissioner Rodney Dillon attended a conference on the *NWI* in April 2004, although the basis for his attendance and the nature and extent of his contribution is unclear.[97]

However, given the fact that the *NWI* Discussion Paper was not released until 19 March 2004 and that key stakeholder meetings on the Discussion Paper were held only between 30 March and 7 April, and only in the state capitals and Canberra, any consultation with Indigenous people on the *NWI* (if any) is likely to have been minimal.

It would appear that whatever consultation occurred with Indigenous people, it translated into only limited recognition of their interests in the resultant *NWI*.

The main clauses in the *NWI* relating to Indigenous interests in water are clauses 52–54. Clause 52 calls for Indigenous involvement in water planning, which is qualified by the phrase 'wherever possible', so it is only aspirational in its terms. Clauses 53 and 54 both refer to native title. The next chapter examines whether native title is capable of meeting Indigenous aspirations for water management, so it is not necessary to go into any detail at this point. However, the conclusion reached in that regard is that native title is an inadequate mechanism to achieve Indigenous aspirations, and therefore the focus on native title in the *NWI* can be seen as a major weakness.

Another relevant clause is clause 25(ix), which is worth setting out as it refers to water management:

> 25. The Parties agree that, once initiated, their ***water access entitlements*** and planning frameworks will: …
> (ix) recognise indigenous needs in relation to water access and management …

Schedule E cl 1(vi), and schedule G cl 3(v) are also relevant to Indigenous interests. Finally, Indigenous and cultural values fall within the definition of 'environmental and other public benefit outcomes'.[98]

with Regard to National Water Initiative Discussion Paper, April 2004, 4, 13; Australian Conservation Foundation, *Submission in Response to the Discussion Paper, National Water Initiative*, April 2004, 16. Copies on file with the author.

96 COAG, *Communiqué* (29 August 2003) 3 (emphasis added).

97 Aboriginal and Torres Strait Islander Commission, *Annual Report 2003–2004* (22 October 2004) 46.

98 *NWI* sch B(i) (definition of 'environmental and other public benefit outcomes').

As a policy document, the *NWI* is not enforceable, and therefore cannot mandate a particular course of action.[99] Nonetheless, it does signify a shift towards the recognition of Indigenous interests in water, a shift that is reflected in the *Water Act 2007* (Cth) ('*Water Act 2007*').

The Water Act 2007 *(Cth)*

As noted in the previous chapter, the *Water Act 2007* also recognises Indigenous interests in water.[100]

For example, pursuant to s 202, the Murray–Darling Basin Authority ('MDBA') must set up a Basin Community Committee ('BCC'), the membership of which must include 'at least 2 Indigenous persons with expertise in Indigenous matters relevant to the Basin's water resources'.[101] The BCC must also establish an Indigenous water subcommittee to 'guide the consideration of Indigenous matters relevant to the Basin's water resources'.[102]

However, this was not initially the case. The original *Water Act 2007* contained a provision enabling a person with expertise in Indigenous matters to be on the BCC, but it was discretionary and contained no requirement for an Indigenous subcommittee. These requirements were inserted pursuant to amendments made in 2008. Notably, the person with Indigenous expertise did not have to be Indigenous.

The limited reference to Indigenous interests in the initial *Water Act 2007* suggests that there was little meaningful, if any, consultation with Indigenous people in relation to its development.[103]

The *Water Act 2007* was the subject of an independent statutory review in 2014.[104] Of the 74 submissions received, 13 referred to the recognition of Indigenous interests.[105] Consultations were also held with the Federation of Victorian Traditional Owner Corporations, the Murray and Lower Darling Rivers Indigenous Nations, the Northern Basin Aboriginal Nations and the National Native Title Council.[106] The review panel released its report in November 2014 and included in its recommendations a number of suggestions arising out of the submissions.[107]

99 Gardner, Bartlett and Gray, above n 82, 52.
100 See, eg, *Water Act 2007* s 21.
101 Ibid s 202(5)(c).
102 Ibid s 202(3)(c).
103 Senator Rachel Siewert noted a lack of consultation with Indigenous stakeholders in the development of the Water Bill 2007: Commonwealth, *Parliamentary Debates*, Senate, 16 August 2007, 250–1 (Rachel Siewert).
104 Section 253 of the *Water Act 2007* requires a review be conducted before the end of 2014.
105 Copies of the submissions are available from <www.agriculture.gov.au/water/policy/legislation/water-act-review>.
106 Expert Panel, *Report of the Independent Review of the Water Act 2007* (November 2014) Appendix C: List of Attendees at Consultations.
107 Ibid, eg, Recommendation 1 (Northern Basin Aboriginal Nations, Submission No 72 to the Expert Panel, *Independent Review of the Water Act 2007*, 30 July 2014, [13]),

So, in this instance, the submissions and consultations clearly had an effect on the review panel.

Impacting on the seriousness with which the review panel considered the submissions and views put forward during the consultations is likely to have been the existence of the legislative requirements in the *Water Act 2007*. Also likely to have been of influence were the Indigenous consultations on the preparation of the Murray–Darling Basin Plan, prompted by the Indigenous requirements of the Basin Plan in the *Water Act 2007*. [108] This lends support to an argument for having legislative acknowledgement of Indigenous interests.

The Australian government proceeded to adopt all the recommendations arising out of the review (two in part), including a recommendation requiring that 'the Act be amended to provide that the Basin Community Committee's membership must include at least two individuals with expertise in Indigenous matters relevant to Basin water resources'.[109] Significantly, though, the government went even further than the recommendation, by requiring that those two individuals be Indigenous.[110]

Concluding remarks

This chapter has demonstrated that Indigenous participation in the development of water management legislation has been fragmented and perhaps even illusory. The historical devaluation of Indigenous citizenship and thus participation could be understood in light of inherent racism and discriminatory attitudes of times past. Less clear is the desultory attention Indigenous interest holders have been accorded in the last 20 years, or what we might recognise as the 'post-*Mabo*' era. The expectations created by *Mabo* that Indigenous voices would be taken seriously are only just starting to be realised in relation to water management legislation. *Mabo* did, however, result in the enactment of the *NTA*, which provides for the recognition of native title rights and interests in water. It is this recognition which is further explored in the next chapter.

and Recommendation 20 (Katie O'Bryan, Submission No 32 to the Expert Panel, *Independent Review of the Water Act 2007*, 4 July 2014, 2, 3).

108 See Chapter 5.

109 Expert Panel, *Report of the Independent Review of the Water Act 2007* (November 2014) Recommendation 20.

110 *Report of the Independent Review of the Water Act 2007 Australian Government Response* (nd) 13.

Part C

Current status of Indigenous water management rights

Commonwealth legislation

4 Native title as a source of Indigenous water rights

Background to the *Native Title Act 1993* (Cth)

Following the seminal 1992 High Court decision in *Mabo v Queensland [No 2]* ('*Mabo*'),[1] the legal fiction that Australia was *terra nullius*, or uninhabited, was finally put to rest. Prior to *Mabo*, in Australian law Indigenous Australians had no legal rights to land arising from their original occupation. The *Mabo* decision thus marked a watershed moment in Australia's history of settler relations with its original Indigenous inhabitants. With the rejection of the *terra nullius* doctrine came the recognition that Australia's Indigenous peoples were the prior owners and occupants of this country, and that any rights and interests that survived British sovereignty could be recognised by the common law. Native title thus emerged, albeit not unscathed, from the shadows of the land ownership regime imported by the colonial settlers.

Mabo was a direct challenge to conventional understandings of Australia's land management and property rights regimes, the very existence of which had been predicated on the assumption that there were no Indigenous rights to land. It was, therefore, imperative that a mechanism be put in place to deal with native title claims, and with activities occurring on land on which native title exists or may be found to exist in the future. Thus, the *Native Title Act 1993* (Cth) ('*NTA*') was born.

The *NTA* was the culmination of intense negotiations involving Indigenous representatives, the Prime Minister, members of the opposition, the Democrats, the Greens, state premiers, and other major stakeholders; including representatives from the mining, pastoralism and farming industries.[2] After what was at the time one of the longest parliamentary sittings in Australia's history, the Native Title Bill 1993 (Cth) was passed.[3] Following a change in government and the High Court decision in

1 (1992) 175 CLR 1.

2 For more detail, see, eg, Maureen Tehan, 'A Hope Disillusioned, an Opportunity Lost? Reflections on Common Law Native Title and Ten Years of the *Native Title Act*' (2003) 27 *Melbourne University Law Review* 523, 538–43; Daryl Cronin, 'The Lead Up to the Passage of the *Native Title Act*' in Toni Bauman and Lydia Glick (eds), *The Limits of Change: Mabo and Native Title 20 Years On* (AIATSIS Research Publications, 2012) 47.

3 Lisa Strelein, *Compromised Jurisprudence. Native Title Cases Since Mabo* (Aboriginal Studies Press, 2006) 4; Commonwealth, *Parliamentary Debates*, Senate, 21 December 1993, 5500 (Gareth Evans).

Wik v Queensland ('*Wik*'),[4] the *NTA* was the subject of major amendments in 1998, with a number of those amendments affecting Indigenous water rights.[5]

By enacting the *NTA*, the Australian Parliament acted in purported recognition of international standards for the protection of human rights, in particular those of Indigenous peoples, and in purported compliance with Australia's obligations under international law.[6] This was certainly the intent of the government when it introduced the Bill.[7]

This chapter identifies those provisions of the *NTA* that recognise Indigenous rights to participate in the management of water resources and evaluates their effectiveness in light of the relevant case law.

The *NTA* and native title rights to water

Relationship between the NTA *and the common law*

The result of *Mabo* was that the common law could recognise those native title rights and interests that had survived the acquisition of sovereignty by the British Crown, provided that such recognition did not 'fracture a skeletal principle of our legal system'.[8] The *NTA* then established a statutory regime for the recognition of native title. This means that it is now the *NTA* rather than the common law to which Indigenous Australians turn if they wish to seek recognition in the courts of their native title rights and interests.[9] The common law and *Mabo* remain relevant, although only to a limited extent.[10]

Mabo referred only to recognition by the common law of native title rights to land. This does not, however, preclude common law recognition of native title rights to waters.[11]

Rulings in subsequent cases have recognised native title rights to waters pursuant to the *NTA* definition of native title. As recognition under the *NTA* can only occur if the rights are 'recognised by the common law of Australia',[12] it is

4 (1996) 187 CLR 1.
5 See, eg, *NTA* ss 24HA, 24MB, 44H.
6 Ibid preamble.
7 Commonwealth, *Parliamentary Debates*, House of Representatives, 16 November 1993, 2878 (Paul Keating, Prime Minister).
8 *Mabo* (1992) 175 CLR 1, 43 (Brennan J).
9 *Commonwealth v Yarmirr* (2001) 208 CLR 1, 35 (Gleeson CJ, Gaudron, Gummow and Hayne JJ) ('*Yarmirr*'); *Yorta Yorta v Victoria* (2002) 214 CLR 422, 440 ('*Yorta Yorta*').
10 They remain relevant insofar as they assist in the interpretation of the *NTA*: see *Western Australia v Ward* (2002) 213 CLR 1, 65–6, 69 (Gleeson CJ, Gaudron, Gummow and Hayne JJ) ('*Ward*'); *Yarmirr* (2001) 208 CLR 1, 111–12 (Kirby J); *Wilson v Anderson* (2002) 213 CLR 401, 456–7 (Kirby J); *Yorta Yorta* (2002) 214 CLR 422, 440–1, 452–4 (Gleeson CJ, Gummow and Hayne JJ). See also Richard Bartlett, *Native Title in Australia* (LexisNexis Butterworths, 3rd edn, 2014) 155.
11 *Yarmirr* (2001) 208 CLR 1, 117 (Kirby J).
12 *NTA* s 253(1)(c).

clear from these cases that the common law recognises at least some form of native title right to waters.

The *NTA* clearly encompasses native title rights in relation to water, which is evident from the long title of the *NTA*. [13] The definition of native title also includes references to waters,[14] which is then specifically defined as including:

> (a) sea, a river, a lake, a tidal inlet, a bay, an estuary, a harbour or subterranean waters; or
> (b) the bed or subsoil under, or airspace over, any waters (including waters mentioned in paragraph (a)); or
> (c) the shore, or subsoil under or airspace over the shore, between high water and low water.[15]

Many determinations of native title provide their own definition of waters, usually of a less inclusive nature.[16] In addition, some determinations use the term 'water' rather than 'waters' without providing a definition, and therefore the ordinary meaning of the term must be used to work out what native title rights and interests are encompassed by the determination.[17] A separate definition of water also exists for the purposes of s 24HA. The reasoning behind having separate definitions and the discussion of native title determinations appear later in this chapter.

Various sections of the *NTA* relate to or have an impact on Indigenous water management rights. The following is an analysis of the most significant, commencing with s 24HA.

Section 24HA – management of water and airspace

Section 24HA is one of the 'future act' provisions of the *NTA*. [18] It is of particular relevance because it relates to the management of water, the role of native title holders and claimants, and notification/consultation procedures. This section was one of the 1998 amendments to the *NTA*, implementing Point 8 of Prime

13 The long title of the *NTA* states that it is '[a]n Act about native title in relation to land or *waters*, and for related purposes' (emphasis added).
14 *NTA* s 223 (1).
15 Ibid s 253 (definition of 'waters').
16 See, eg, *Nangkiriny v Western Australia* (2002) 117 FCR 6, 12 (North J); *Mervyn v Western Australia* [2005] FCA 831 (29 June 2005).
17 Katie O'Bryan, 'Issues in Natural Resource Management – Inland Water Resources – Implications of Native Title and the Future of Indigenous Control and Management of Inland Waters' (2007) 14(2) *Murdoch University E-law Journal* 280, 290.
18 The term 'future act' is defined in s 233 of the *NTA*. It does not include 'past acts', defined in s 228. The operative future act provisions, contained in pt 2 div 3, constitute a regime by which 'future dealings affecting native title may proceed' and 'set[s] standards for those dealings': at s 3(b).

Minister John Howard's 'Wik 10 Point Plan',[19] the government's response to the so-called 'uncertainty' it saw as having been created by the *Wik* decision.[20]

The term 'water', for the purposes of this section, means 'water in all its forms'. This definition deliberately excludes the bed or subsoil under, or airspace over, any waters. The Explanatory Memorandum to the Native Title Amendment Bill 1997 (Cth) ('Explanatory Memorandum') explains the purpose for this exclusion as follows:

> Subdivision H does **not** deal with the management and regulation of the bed or subsoil under onshore and offshore waters (which includes the regulation of off-shore mining such as the petroleum and gas industries). These matters are generally dealt with in proposed Subdivisions M and N, respectively, of Division 3 (see Chapters 15 and 16). It is for this reason that the term 'water' is used rather than the term 'waters' (which is defined in section 253 to include the bed or subsoil).[21]

The Explanatory Memorandum then explains the effect of s 24HA as being to ensure

> that legislation and other future acts dealing with surface and sub-surface water, living aquatic resources and airspace will be valid. Any rights or obligations created by the acts will prevail over inconsistent native title rights and interests. The non-extinguishment principle applies to these acts and native title holders are entitled to compensation from the relevant government.[22]

Under s 24HA, the making, amendment or repeal of legislation or the grant of a lease, licence, permit or authority that relates to the management or regulation of water is valid. The non-extinguishment principle applies, and compensation is payable for any effect the act may have on native title rights.

Unlike future acts which relate to mining,[23] future acts under this section do not attract the right to negotiate. Instead, they fall within the class of future acts to which only the rights to be notified and to have an opportunity to comment apply.[24] These rights do not, however, apply to the making, amending or repeal of legislation.

In relation to the right to be notified and the opportunity to comment, those rights are very limited in scope, as the following cases demonstrate.

19 John Howard, 'Amended Wik 10 Point Plan' (Press Release, 8 May 1997) <http://pmtranscripts.pmc.gov.au/release/transcript-10333>.
20 John Howard, 'Transcript of the Prime Minister the Hon John Howard MP Press Conference – Parliament House' (7 May 1997) <https://pmtranscripts.pmc.gov.au/release/transcript-10331>.
21 Explanatory Memorandum, Native Title Amendment Bill 1997 (Cth) [10.2].
22 Ibid [10.1].
23 *NTA* sub-div P.
24 Ibid s 24HA(7)(b).

The scope of the opportunity to comment was discussed in *Harris v Great Barrier Reef Marine Park Authority* ('*Harris*').[25] In *Harris*, the Full Federal Court stated:

> The right under s 24HA(7)(b) is, we think, a right to proffer information and argument to the decision-maker that it can make such use of as it considers appropriate. The sub-section does not confer any greater right on the native title interests. It is not a right to participate in the decision whether to issue the permit or a right that entitles the recipients to seek information from the decision-maker necessary to satisfy those interests about matters of concern to them.[26]

The Full Court made a clear distinction here between the right to comment, as found in s 24HA, and other rights,[27] such as those found in other parts of pt 2 div 3: the right to make submissions,[28] the right to object,[29] the right to be consulted[30] and the right to negotiate.[31] The Full Court was clearly of the view that the right to comment was a lesser one and that if the legislature had intended it to be more substantial, it would have provided as such.

In relation to notification, the applicants in *Harris* argued that the common law requirements of procedural fairness entitled them to be provided with sufficient information to enable them to have

> a proper opportunity to advance all legitimate arguments to avert a decision that might profoundly affect their interests, an opportunity that in turn requires that they be given a proper notice of the case they have to meet.[32]

This argument was rejected by the court on the basis that the *NTA* provided a statutory regime to meet the requirements of procedural fairness.[33]

In relation to the content of the notification, the Federal Court stated:

> The ordinary meaning of the phrase 'or acts of that class' in s 24HA(7) suggests that it is not necessary for the Authority to give notification to the registered native title claimants that it is proposing to grant each specific permit of a class of permit proposed to be granted. That that is also the intended meaning is confirmed by par 10.20 of the explanatory memorandum.[34]

25 (2000) 98 FCR 60.
26 Ibid 71.
27 See ibid 71–4.
28 See, eg, *NTA* s 31(1).
29 See, eg, ibid ss 24CI(1), 24DJ(1).
30 See, eg, ibid s 24JAA(13)–(15).
31 Ibid pt 2 div 3 sub-div P.
32 *Harris* (2000) 98 FCR 60, 65.
33 Ibid 67–8.
34 Ibid 72.

Thus, the Full Federal Court in *Harris* gave a narrow scope to both the opportunity to comment and notification requirements in s 24HA(7).

In a subsequent decision of the National Native Title Tribunal ('NNTT'), Member Sosso saw fit to be guided by *Harris* on the adequacy of notice.[35] However, there appears to have been no further cases that specifically deal with the opportunity to comment.

A failure to notify does not render the act invalid. This was clearly the intention of the government.[36] Although there has been no definitive judgment on this point, it was the view expressed in obiter by a majority of the Full Federal Court in *Lardil Peoples v Queensland* ('*Lardil*').[37] As Dowsett J stated:

> Section 24 HA(3) validates a future act without any suggestion that such validation is dependent upon compliance with any other aspect of the section or subdivision. Native title rights are not extinguished by the act in question, but enjoyment of them may be suspended. Compensation is payable, but there is no suggestion that payment is a condition of validity. There is also nothing to suggest that compliance with s 24HA(7) is a condition precedent to validity.[38]

This reasoning conforms with the intention of the government, but it renders any efforts by native title claimants and holders to have their comments considered in the decision-making process largely futile. In other words, even if a court held that a decision-maker had failed to follow the notification procedures in s 24HA(7) and had granted the licence, there would be little, if any, value for native title holders to then provide comments: the licence has been granted and is valid. There is also no incentive for the decision-maker to follow the notification procedures due to the absence of a legal sanction or consequence for non-compliance. The only disincentive might be a loss of goodwill from the affected native title holders.

Bartlett has suggested that the view of the Full Federal Court in *Lardil* is incorrect.[39] In support, he refers to the reasoning of the Full Federal Court in *Harris*, stating that '[i]n explaining the procedural entitlement of claimants, the court assumed that non-compliance led to invalidity'.[40] In that regard, the Full Court made various observations of procedural rights having the effect of enabling claimants to be involved in the validation process, of which one observation is particularly pertinent:

> The non-extinguishment principle, defined in s 238 applies to most of the future acts covered by Subdivisions G, H, I and J. Though the native title holders

35 *Dann and Western Australia and Empire Oil Co (WA) Ltd* [2006] NNTTA 126 (25 August 2006) [58].
36 Explanatory Memorandum, above n 21, [10.20].
37 (2001) 108 FCR 453, 471–2, 473 (French J), 486–7 (Dowsett J).
38 Ibid 486.
39 Bartlett, *Native Title in Australia*, above n 10, 591.
40 Ibid.

> cannot prevent the doing of these future acts, they … have certain procedural rights to which effect must be given before the act can validly be done.[41]

Nonetheless, several cases have taken guidance from the view of the court in *Lardil* on the issue of validity. In *Daniel v Western Australia* ('*Daniel*'),[42] Nicholson J, after noting the dicta of the court in *Lardil*, agreed with a respondent's submission that there was 'no reason to depart from such persuasive authority which is not clearly incorrect'.[43] In *Banjima People v Western Australia [No 2]*, Barker J also accepted respondent submissions on *Lardil* and noted *Daniel.* [44] And, in *Queensland Construction Materials Pty Ltd v Redland City Council*, Chesterman JA and Applegarth J, although noting that the decision in *Lardil* was not binding, were clear that they would be reluctant to depart from that decision unless clearly wrong, and here they thought it was 'plainly right'.[45] They then proceeded to cite, with apparent approval, French J in *Lardil* in relation to this particular issue (even though it was not necessary for the appeal).[46]

Given these cases, and the clearly stated intention of the Explanatory Memorandum, the likelihood of a later court now coming to a different conclusion appears remote.[47]

To conclude, under s 24HA native title holders and claimants have no rights to be notified of or given the opportunity to comment on proposed legislation about the management or regulation of water. In relation to future acts whereby such rights exist, the scope of the right to comment is narrow, notification requirements are minimal, and the act remains valid whether or not notification has occurred. No right is available under this section to participate in or even be consulted about decisions regarding the management of water resources. The most that native title holders can expect from this section is a right to compensation should their native title rights to water be affected.

Indigenous Land Use Agreements

Introduced in 1998, the provisions relating to indigenous land use agreements ('ILUAs') were 'generally seen to be a positive element of the 1998 package'.[48] They were intended to enable agreements to be reached between 'actual or potential native title holders and those wishing to use land'.[49] If such an

41 *Harris* (2000) 98 FCR 60, 68–9.
42 (2004) 212 ALR 51.
43 Ibid 65. *Daniel* was upheld in *Moses v Western Australia* (2007) 160 FCR 148.
44 *Banjima People v Western Australia [No 2]* (2013) 305 ALR 1, 165.
45 (2010) 175 LGERA 52, 73.
46 Ibid 73, 74.
47 See also *Western Desert Lands Aboriginal Corporation v Western Australia* (2008) 218 FLR 362, 379 (Deputy President Sumner); *Leach v Nominal Defendant (QBE Insurance (Australia) Ltd) [No 2]* [2014] NSWCA 391 (14 November 2014) (McColl JA).
48 Strelein, above n 3, 8.
49 Explanatory Memorandum, above n 21, [6.7].

agreement is in place, 'then its terms are intended to take precedence over any other provisions in the *NTA* which would otherwise apply to the future acts covered by the agreement'.[50] Agreement making is seen as beneficial because it avoids litigation and

> is an important process through which people build relationships and carry forward the public recognition of Indigenous rights ... [and] to varying degrees, gives Indigenous people a genuine decision making role in a range of issues affecting their lives and their territories.[51]

The potential content of ILUAs is extremely wide. Provided an ILUA includes one or more of the native title matters enumerated in the relevant section, it remains free of any other limitations.[52] An exception relates only to alternative procedure ILUAs, which are not able to provide for the extinguishment of native title.[53] At the time of writing, no alternative procedure ILUAs were in place.

Given the wide-ranging scope of ILUA content, it is conceivable that an ILUA could provide for rights to participate in the management of water resources.

However, a question arises as to how rights contained in ILUAs interact with the rights contained in s 24HA (and other future act provisions), given the potentially wide scope of the content of ILUAs. This is answered by s 24AB, which states:

> **24AB Order of application of provisions**
>
> Indigenous land use agreement provisions
>
> (1) To the extent that a future act is covered by section 24EB (which deals with the effect of indigenous land use agreements on future acts), it is not covered by any of the sections listed in paragraphs 24AA(4)(a) to (k).

What this section means is that the validity of the future act depends on compliance with the procedural rights contained in the ILUA rather than the procedural rights contained in the sections listed in s 24AA(4)(a) to (k), of which s 24HA is one. Once registered, ILUAs are binding as a contract between parties;[54] this means that parties are essentially 'contracting out' of the future act provisions of the *NTA* and can negotiate their own procedures for dealing with particular matters. Theoretically, a native title party could negotiate for inclusion in an ILUA of procedural rights in relation to the management of water resources that are greater than those contained in s 24HA.

50 Ibid.

51 Marcia Langton, Maureen Tehan and Lisa Palmer, 'Introduction' in Marcia Langton et al (eds), *Honour among Nations? Treaties and Agreements with Indigenous People* (Melbourne University Press, 2004) 25.

52 *NTA* ss 24BE, 24CE and 24DF.

53 Ibid s 24DC.

54 *NTA* s 24EA.

In practice, this is unlikely to occur, given the relative bargaining positions of native title parties vis-à-vis non-Indigenous parties.[55] There are no incentives or benefits (other than perhaps increased goodwill) for states or other parties to do more than is required under s 24HA.

Most ILUAs are privately reached between the parties concerned, with only limited details made publicly available.[56] Accordingly, it is difficult to ascertain the effectiveness of ILUAs to provide for water rights greater than those contained in s 24HA.

Future act provisions – subdivisions M and P

The definition of water in s 24HA deliberately excludes the bed or subsoil under waters so that they can be dealt with under other future act provisions of the *NTA*. Part 2, div 3, sub-div M is relevant in this regard.[57] Also of relevance is pt 2, div 3, sub-div P. For the purposes of brevity, these subdivisions will be referred to as 'sub-div M', and 'sub-div P' respectively.

Subdivision M – treatment of acts that pass the freehold test

Subdivision M relates to acts passing the freehold test.[58] The basic concept behind this test is that 'for the purposes of providing equality before the law, future acts should only be valid over native title lands or waters if they could also be done over freehold and subject to similar conditions and procedural requirements'.[59] As it follows s 24HA, the terms of s 24HA take priority.[60]

Under sub-div M, native title holders are entitled to the same procedural rights as if they held freehold title.[61] Thus, if a freehold owner of land has a right to be consulted or to lodge an objection, then native title holders would also have those rights. Similarly, if a freehold owner only has the right to be notified and given the opportunity to comment, then native title holders would face the same restrictions. However, if the act is one to which sub-div P applies (the 'right to negotiate'), then those procedural rights apply rather than the procedural rights in sub-div M.[62]

At first glance, it is difficult to reconcile sub-div M with the priority given to s 24HA. Yet, if one recalls that the definition of waters which applies to s 24HA

55 See Langton, Tehan and Palmer, above n 51, 20.
56 The NNTT maintains a register of ILUAs pursuant to *NTA* s 199A. The content of the register is limited to only those details of the ILUA as set out in s 199B; not the entire ILUA.
57 Explanatory Memorandum, above n 21 [10.2].
58 *NTA* s 24MB.
59 Bartlett, *Native Title in Australia*, above n 10, 616.
60 *NTA* s 24AB(2).
61 Ibid s 24MD(6A).
62 Ibid s 24MD(6).

does not include the bed or subsoil under waters, then to those elements, at least, the freehold test in sub-div M would still apply.

Subdivision P – right to negotiate

Subdivision P contains some of the strongest provisions for native title claimants and holders. Under this subdivision, registered native title claimants and holders have a right to negotiate on certain future acts occurring on their traditional land, such as the conferral of mining rights or compulsory acquisitions.[63] By far the majority of right-to-negotiate matters relate to the conferral of mining rights. Mining operations use a significant amount of water, so this subdivision is of particular importance in that regard. Pursuant to this subdivision, the parties must negotiate 'with a view to reaching an agreement about the act'.[64] There are notification requirements to be met, which include informing native title holders and registered claimants, plus the provision of prescribed documents and information.[65]

After the negotiating parties have been identified, the procedure includes a requirement that they negotiate in good faith in an effort to obtain the agreement of the native title parties.[66] But native title parties have no right of veto; if the parties are unable to reach an agreement within six months, then the matter may be taken to arbitration.[67] If this is the case, it is unlikely to be resolved in favour of the native title party,[68] and conditions imposed on the non-native title party (if any) are usually minimal.[69]

The scope of negotiations and the content of an agreement are not specifically limited by the terms of the *NTA*, although certain matters are able to be included.[70] But if agreement cannot be reached about 'matters that are unrelated to the effect of the act on the native title rights and interests of the native title parties, this does not mean that the negotiation party has not negotiated in good faith'.[71] This can be a limitation because a party can seek arbitration after six months of negotiation if it has negotiated in good faith only on 'matters related to the effect of the act on the native title rights and interests of the native title parties', even if it

63 Ibid s 25(1)(a).
64 Ibid s 25(2).
65 Ibid s 29(4).
66 Ibid s 31(1)(b).
67 Ibid s 35.
68 As at 21 January 2018 there had been 103 future act determinations on whether an act can proceed or not (excluding those determinations made by consent or dismissed). Of those 103, only three resulted in a determination that the act must not be done. Of the remaining 100 determinations, 59 resulted in a determination that the act can be done, and 41 resulted in a determination that the act can be done with conditions.
69 See, eg, *Cheedy v Western Australia* [2012] NNTTA 11 (7 February 2012) [63] (Member O'Dea).
70 *NTA* s 33.
71 Ibid s 32(2).

has refused to negotiate in good faith on other aspects of importance to the native title parties.

Under this subdivision, native title parties during negotiations can clearly discuss the impact of, for example, proposed mining activity on their native title rights to waters. As well as this initial input, they may also be able to negotiate as a term of the agreement that there be continuing consultations or some other form of involvement in decision-making in relation to mining activities that could affect their native title rights, including water rights, on the land that is the subject of the agreement. This could be by way of representation on an environment committee (if the company has one), or by the establishment of a liaison committee to monitor the implementation of the agreement. However, a refusal by a party to negotiate such terms to be included in an agreement may not prevent a matter from going to arbitration as they may be seen as not directly relating to 'the effect of the act on the native title rights and interests'. Thus, it will largely depend on the views of the non-native title party on the value of having (or maintaining) a good relationship with the native title party as to whether it is willing to include terms relating to matters going beyond simply those that affect native title rights and interests.

Unfortunately, agreements reached pursuant to sub-div P are not publicly available, generally being in the nature of a private contract between the parties concerned. Further, the *NTA* does not require that the NNTT maintain a register of sub-div P agreements. Given these limitations, as with ILUAs, it is difficult to ascertain the effectiveness of sub-div P agreements with respect to water rights, beyond what the *NTA* specifically elicits.

Other future act provisions relating to native title rights to inland waters

Various other sections of the *NTA* relate to future acts and water rights.[72] As these sections do not specifically relate to the right to participate in the management of water resources, it suffices to note that they generally provide for a right to be notified and an opportunity to comment. As the content of the right to be notified and scope of the opportunity to comment were found in *Harris* to be very limited, these future act provisions of the *NTA* clearly provide little protection of Indigenous water rights.

General provisions of the NTA *affecting native title rights to inland waters*

Section 44H – rights conferred by valid leases, etc.

Section 44H is also relevant; it, too, was introduced as part of the 1998 amendments. In summary, s 44H states that any activities performed in accordance with

72 *NTA* s 24GA–24GE, s 24IA–24ID, s 24IA–24ID. For a more detailed discussion of these sections of the *NTA*, see Lila D'souza, 'Native Title Implications for Existing and Future Water Entitlements in Western Australia' (Legal Report, National Native Title Tribunal, 3 May 2002).

a lease, licence, permit or authority are valid, take priority over native title rights and interests but do not extinguish them, and no further compensation is payable with respect to those activities. Moreover, no requirement exists to notify native title holders of these activities. Thus, despite the fact that they may affect native title rights and interests, native title holders have no rights to have any input into how to minimise or even prevent the activity's potentially harmful effects on native title, including native title rights to water. An example of such an activity would be the construction of a dam under a valid pastoral lease.[73]

It is therefore important for native title holders and claimants to ensure that they are involved in the initial grant of the lease, licence or permit as no opportunity will be available at a later stage to be consulted about activities conducted under that lease, licence or permit. Nor will native title holders and claimants be able to obtain compensation for the effect on native title by those activities at that later stage.

But given that procedures relating to the initial grant of a lease, licence, permit or authority may in many cases only provide minimal notification rights and a very limited opportunity to comment, it seems that this section merely compounds the lack of protection of Indigenous water rights.

Section 211 – preservation of certain native title rights and interests

Section 211 of the *NTA* provides that where a law of a state or territory would normally prohibit or restrict a class of activity (defined in s 211 as hunting, fishing, gathering and conducting cultural or spiritual activities) other than in accordance with the relevant permission, the rights of native title holders to undertake those activities and to access land and waters for the purposes of undertaking those activities is preserved. This is subject to some qualification, as the activities undertaken must be:

(a) for the purpose of satisfying their personal, domestic or non-commercial communal needs; and
(b) in exercise or enjoyment of their native title rights and interests.[74]

In addition, native title holders are still subject to laws of general application (for example, if they propose to go hunting with a gun, they will require a gun permit). It is also arguable that the activity must not have been prohibited outright, merely regulated.[75]

73 Explanatory Memorandum, above n 21, [6.21].
74 *NTA* s 211(4).
75 D'souza, above n 72, 29. See also *Yanner v Eaton* (1999) 201 CLR 351, 372–3 (Gleeson CJ, Gaudron, Kirby and Hayne JJ); *Ward* (2002) 213 CLR 1, 152 (Gleeson CJ, Gaudron, Gummow and Hayne JJ); *Neowarra v Western Australia* [2003] FCA 1402 (8 December 2003) [633], [636] (Sundberg J); *Karpany v Dietman* (2013 (2013) 252 CLR 507, 514.

Section 211 relates to water rights in that it preserves access to waters, and the conducting of cultural and spiritual activities insofar as those activities relate to water. It could be argued that management, or aspects thereof, of water resources is both a cultural and spiritual activity and would thus fall within the class of activities covered by this section. However, how this kind of activity operates within a regime of granting permissions is unclear; it could, perhaps, include activities such as diverting water or the removal of vegetation from river banks.

So this section may have some scope, albeit limited, for the continuance of some cultural and spiritual activities that have an impact on water management. There has, however, been no judicial consideration of this point.

Section 212 – confirmation of ownership of natural resources, access to beaches, etc.

Section 212 enables the Commonwealth, states and territories to pass legislation confirming the Crown's existing ownership of natural resources; its right to use, control and regulate the flow of water; fishing access rights, and existing public access to waterways, beaches and other public places. Such confirmation does not extinguish native title.[76]

All states and territories have passed such legislation.[77] Confirmation of the Crown's right to use, control and regulate the flow of water has implications for participation in water management by native title holders, particularly given that nearly all the states have water management legislation which vests such rights in the Crown. Case law examining some of this legislation is discussed in the following section.

Judicial consideration of inland water rights

The most significant native title case relating to the recognition of Indigenous water rights is the High Court's 2002 decision in *Western Australia v Ward* ('*Ward*').[78] In *Ward*, the High Court determined that Western Australian legislation vesting the 'right to the use and flow and to the control of the water'[79] in Western Australia did not extinguish native title rights in water, but destroyed any exclusivity.[80] However, prior to *Ward*, in *Wandarang People v Northern Territory*

76 *NTA* s 212(3).

77 *Native Title Act 1994* (ACT) ss 10–13; *Native Title (New South Wales) Act 1994* (NSW) ss 16–18; *Validation (Native Title) Act 1994* (NT) ss 12–13; *Native Title (Queensland) Act 1993* (Qld) ss 16–18A; *Native Title (South Australia) Act 1994* (SA) s 39; *Native Title (Tasmania) Act 1994* (Tas) ss 13–14; *Land Titles Validation Act 1994* (Vic) ss 14–16; *Titles (Validation) and Native Title (Effect of Past Acts) Act 1995* (WA) ss 13–14.

78 (2002) 213 CLR 1.

79 *Rights in Water and Irrigation Act 1914* (WA) s 4(1).

80 This was only a small aspect of *Ward*, but it was very significant in relation to water rights.

(‘*Wandarang*’)[81] Olney J had already briefly considered this issue in relation to Northern Territory legislation.

In *Wandarang*, Olney J opened by referring to common law recognition under s 223(1)(c), stating that ‘[s]ection 223(1)(c) of the *Native Title Act* makes it clear that a right or interest which is not recognised by the common law of Australia cannot be recognised as a native title right or interest’.[82] His Honour then elaborated on the limits of common law recognition, citing both Brennan J in *Mabo* [83] and Kirby J in *Fejo v Northern Territory* [84] with approval.[85]

Olney J later concluded by stating:

> As previously discussed, the common law does not recognise a claim to ownership of flowing water. In relation to water generally, the *Control of Waters Ordinance 1938* (now replaced by the *Water Act 1992*) has established a regime in relation to water rights which is inconsistent with the continued existence of exclusive native title rights to the ownership and use of water.[86]

It is this view of the *Control of Waters Ordinance 1938* (NT) that was later to be largely replicated by the High Court in *Ward* in relation to the equivalent Western Australian legislation.[87]

Ward *and inland water rights*

At first instance, Lee J found that the vesting of the ‘right to the use and flow and to the control of the water’ in pt III of the *Rights in Water and Irrigation Act 1914* (WA) (‘*RWI Act*’) did not extinguish native title rights to water.[88]

On appeal to the Full Federal Court, Western Australia submitted that Lee J had erred in coming to this conclusion. However, the majority of the Full Court did not accept this submission.[89] Nonetheless, the court proceeded to accept Western Australia’s argument that the *RWI Act* removed any exclusive native title rights to water.[90]

The matter then proceeded on appeal to the High Court, which upheld, by majority, the majority view of the Federal Court, stating that ‘[t]he vesting of waters in the Crown was inconsistent with any native title right to possession of those waters to the exclusion of all others’.[91] It is here that we see the similarities with the judgment of Olney J in *Wandarang*.

81 (2000) 104 FCR 380.
82 Ibid 392.
83 (1992) 175 CLR 1.
84 (1998) 195 CLR 96.
85 *Wandarang* (2000) 140 FCR 380, 392.
86 Ibid 432.
87 Note that the High Court in *Ward* did not refer to *Wandarang*.
88 *Ward v Western Australia* (1999) 159 ALR 483, 582–3.
89 *Western Australia v Ward* (2000) 99 FCR 316, 422 (Beaumont and von Doussa JJ).
90 Ibid 423 (Beaumont and von Doussa JJ).
91 (2002) 213 CLR 1, 152.

Comparable vesting provisions are found in most other states and territories in Australia.[92] In Victoria, for example, s 7(1) of the *Water Act 1989* (Vic) ('*Water Act 1989*') provides that '[t]he Crown has the right to the use, flow and control of all water in a waterway and all groundwater'. Thus, it is clear that, almost Australia-wide, any native title rights to water recognised by the courts will be non-exclusive.[93]

At this juncture, mention should be made of the High Court's 2009 decision in *ICM Agriculture Pty Ltd v Commonwealth.* [94] In that case, the High Court stated that common law riparian rights to water in New South Wales had been extinguished by statutory vesting provisions,[95] and noted similar historical vesting provisions around Australia.[96] However, native title rights, although they can be recognised by the common law, are not common law rights.[97] It has been argued that this distinction is 'an important distinction, as it shields native title rights to use water from any extinguishment that may have occurred to common law riparian rights as a result of the universal vesting of water in the Crown in Australia'.[98]

This argument has not yet been tested in court, and, to date, native title determinations continue to be made around Australia which recognise non-exclusive native title rights to waters.

Other native title cases discussing water rights

The High Court decision in *Ward* relating to the effect of the *RWI* Act was applied in *Daniel v Western Australia* [99] and *Neowarra v Western Australia.* [100]

In the 2006 case of *Griffiths v Northern Territory* involving (among other things) the Northern Territory's *Water Act 1992*, and its predecessor, the *Control*

92 *Water Resources Act 2007* (ACT) s 7; *Water Management Act 2000* (NSW) s 392; *Water Act 1992* (NT) s 9; *Water Act 2000* (Qld) s 19; *Water Management Act 1999* (Tas) s 7; *Water Act 1989* (Vic) s 7. South Australia is the only state which does not have such vesting provisions.

93 South Australia does not have legislation vesting the use, flow and control of water in the state; however, the common law position would probably still find no exclusive possession of water, as this would amount to ownership which the common law will not recognise: see *Wandarang* (2000) 104 FCR 380, 432.

94 (2009) 240 CLR 140.

95 Ibid 177 (French CJ, Gummow and Crennan JJ).

96 Ibid 172–3.

97 *Yorta Yorta* (2002) 214 CLR 422, 453 (Gleeson CJ, Gummow and Hayne JJ). See also *Fejo* (1998) 195 CLR 96, 128 (Gleeson CJ, Gaudron, McHugh, Gummow, Hayne and Callinan JJ).

98 Michael O'Donnell, *NAILSMA – TRaCK Project 6.2: Indigenous Rights in Water in Northern Australia* (Charles Darwin University, 2011) 48. At 32–50, O'Donnell discusses at some length the impact of *ICM Agriculture* (2009) 240 CLR 140 on Indigenous water rights in the three northern Australian jurisdictions.

99 [2003] FCA 666. *Daniel v Western Australia* was upheld by the Full Court in *Moses v Western Australia* (2007) 160 FCR 148.

100 [2003] FCA 1402 (8 December 2003).

of Waters Ordinance 1938, Weinberg J discussed the special position of the waters of Timber Creek.[101] Although some confusion arose as to the nature of the water rights being claimed, his Honour eventually found that the native title rights and interests in water were non-exclusive. In reaching this conclusion, he noted Mansfield J's comments in *Gumana v Northern Territory [No 2]* (on which the Northern Territory had relied in its submission) whereby Mansfield J remarked that although there could be no ownership in or exclusive rights to free-flowing or subterranean waters, there would be a right to control access to waters in areas of exclusive possession.[102] More recently, in *Rrumburriya Borroloola v Northern Territory*, Mansfield J made similar remarks to those he had made in *Gumana [No 2]*.[103] This is potentially very important, and may provide one of the only bases upon which native title holders can assert their rights to make decisions about the management of waters on their native title lands.[104] A finding of exclusive possession, however, is very unlikely in the more settled areas of Australia.

Determinations of native title and water rights – Victoria

Apart from the cases just mentioned, most cases do not specifically discuss native title rights and interests in waters but will simply recognise them (or not, as the case may be) in the relevant determination. The content of native title rights and interests in waters recognised by the courts in such cases will vary between groups, and will depend on a number of interrelated factors.[105] The content of rights and interests in waters has also varied over time, as both claimants and respondents become more familiar with the *NTA* and what is or is not possible to achieve.

In Victoria, for example, there have been five determinations of native title: one litigated and four by way of a consent determination. In chronological order they are *Yorta Yorta v Victoria* ('*Yorta Yorta*'),[106] *Clarke v Victoria* ('*Wimmera Clans*'),[107] *Lovett v Victoria* ('*Gunditjmara*'),[108] *Mullett v Victoria* ('*Gunai/Kurnai*'),[109] and *Lovett v Victoria [No 5]* ('*Gunditjmara and Eastern Maar*').[110] None of them resulted in the recognition of exclusive possession.

101 (2006) 165 FCR 300, 371–3.
102 *Gumana v Northern Territory [No 2]* [2005] FCA 1425 (11 October 2005) [33]–[43].
103 [2016] FCA 776, [461].
104 See also O'Donnell, above n 98, 10, 59–61, 303.
105 These will include: the rights and interests claimed in the original native title application; the location of the native title claim; the types of water sources found within the claim; the negotiating strength of the parties (where the determination is one made by consent); the nature of other interests in the claim area; and policy and political imperatives of the government of the day.
106 (2002) 214 CLR 422.
107 [2005] FCA 1795 (13 December 2005).
108 [2007] FCA 474 (30 March 2007).
109 [2010] FCA 1144 (22 October 2010).
110 [2011] FCA 932 (27 July 2011).

In the *Yorta Yorta* decision it was held that native title did not exist, which means that the Yorta Yorta People cannot rely on native title as a basis for recognising their interests in water.

In *Wimmera Clans*, native title was held to exist along the banks and beds of a section of the Wimmera River, Outlet Creek, and over Lake Albacutya and Lake Hindmarsh ('Determination Area A').[111] However, pursuant to the determination, native title does not exist in 'any waters within Determination Area A. For the avoidance of doubt, waters do not include the bed or subsoil under, or airspace over, any waters within Determination Area A.'[112] Given the definition of waters in the *NTA* and that the determination itself only relates to the banks and beds of various water bodies, without this clarification there would have been effectively no recognition of native title.

The *Wimmera Clans* determination then sets out the native title rights that were recognised, namely 'the non-exclusive rights to hunt, fish, gather and camp for personal, domestic and non-commercial communal needs'.[113] As a result, the peoples of the Wimmera Clans can camp on the banks of the Wimmera River, fish in and gather reeds from the river, but have no native title rights to take or use water from it. Thus, even with a consent determination in their favour, the Wimmera Clans, like the Yorta Yorta, cannot rely on native title as a basis for recognition of their interests in water.

The extremely limited native title rights and complete lack of native title rights to water that were recognised in *Wimmera Clans* can perhaps be explained by two factors.

First, *Wimmera Clans* was the first claim to be determined in Victoria after *Yorta Yorta*. As a result of *Yorta Yorta*, it was a commonly held view that the chances of a positive determination in Victoria, indeed in any part of southern Australia, were almost non-existent.[114]

Second, at the time, Victoria was experiencing a severe drought. Accordingly, reluctance by the Victorian government to agree to recognise any native title rights to water, no matter how limited, would not be difficult to infer. This reluctance was, perhaps, an overreaction given that the nature of native title rights to water that had been recognised elsewhere in Australia was generally limited to domestic, non-commercial communal-type rights. The exercise of native title rights of this nature is likely to have minimal if any impact on water resources. Further, pursuant to s 8 of the *Water Act 1989*, a person may use water from any waterways accessible by public road or reserve for domestic and stock purposes. Given that the determination of native title was over public land, and was therefore publicly accessible, the Victorian government would have been agreeing to

111 [2005] FCA 1795 (13 December 2005) order 3 (Merkel J).
112 Ibid order 4(a).
113 Ibid order 7.
114 See, eg, Richard Bartlett, 'An Obsession with Traditional Laws and Customs Creates Difficulty Establishing Native Title Claims in the South: Yorta Yorta' (2003) 31 *University of Western Australia Law Review* 35.

recognise water rights for the Wimmera Clans arguably no greater than they already had as members of the public.

Wimmera Clans did, however, pave the way for some improvement in the recognition of native title rights in the subsequent *Gunditjmara* and *Gunai/Kurnai* determinations.

The rights and interests recognised are similar in each case, being various non-exclusive rights over land and now also over waters. Rights in relation to water are limited to the right to take water from a waterway for domestic and ordinary use. However, although the term 'waters' is used in each determination, the default *NTA* definition of waters is qualified to exclude groundwater.[115] This qualification is incompatible with both the holistic view of country held by Indigenous people and the objectives of the *National Water Initiative*, one of which is to achieve the recognition of connectivity between surface and groundwater resources.[116] Therefore, the water rights recognised in these determinations remain problematic.

An important point to note about the Victorian determinations is that they pertain to water access and use, rather than management. For a determination to include management rights over water resources, the native title rights claimed would need to include a right to make decisions about the land and waters. Although generally always claimed (and indeed claimed in all the Victorian determinations), this kind of native title right is usually only recognised in determinations of exclusive native title.[117] Such determinations are extremely unlikely to occur in Victoria given its land tenure history, the myriad of other interests in land, the nature of Victorian determinations to date, and the enactment of the *Traditional Owner Settlement Act 2010* (Vic). A similar conclusion could be reached about other areas of settled Australia.

The other native title right claimed that may give rise to a right to manage water resources is the right to protect sacred sites or places of importance, where those sites or places include water resources. But there are limitations on this right as it does not allow for the holistic management of water resources, providing for protection of only the site or place in question. This may be relatively simple if the sacred site or place of importance is a rock hole or similarly confined location, but when it is part of a larger water body, such as a river, it becomes problematic, as it has the potential to fragment the management of that larger water body.

115 *Gunai/Kurnai* [2010] FCA 1144 (22 October 2010) determination order 3(c), *Gunditjmara* [2007] FCA 474 (30 March 2007) determination order 3(c); *Gunditjmara and Eastern Maar* [2011] FCA 932 (27 July 2011) determination order 3(c).

116 *Intergovernmental Agreement on a National Water Initiative* (25 June 2004) cl 23(x).

117 There are some non-exclusive determinations in which a limited right to make decisions has been recognised, but in such cases this right only relates to decisions about the use of the land and waters by other Aboriginal people, and not the wider non-Aboriginal community. See, eg, *Apetyarr v Northern Territory* [2014] FCA 1088 (14 October 2014) order 6(i).

The *NTA* was the subject of an inquiry by the Australian Law Reform Commission ('ALRC'), which released its final report on 4 June 2015.[118] Although the report discussed (among other things) the nature and content of native title rights and interests,[119] it had little to say about water, with the ALRC in that regard making 'no specific recommendation'.[120] The limited discussion centred largely on the current recognition of non-exclusive rights to take and use water for domestic, personal and communal needs, and the potential extension of those rights to commercial uses.[121] No mention was made of amending the *NTA* to improve procedural rights under s 24HA.[122] However, the report did state that there was 'merit in a broader review of native title rights in relation to water'.[123]

In late 2017 the Commonwealth government commenced a public consultation process on options for amending the *NTA*, drawing on recommendations made by a number of reviews of the native system, including the ALRC Report.[124] However none of the options presented for comment relate to the management of water resources.[125]

An alternative process in Victoria now exists for the settlement of native title claims: the *Traditional Owner Settlement Act 2010* (Vic) ('*TOSA*'), which is the Victorian government's preferred process.[126] As of May 2018, only two native title claims had been resolved utilising the *TOSA*, namely Gunai/Kurnai and Dja Dja Wurrung. The Dja Dja Wurrung native title claim was settled under the *TOSA* without a native title outcome,[127] and thus the Dja Dja Wurrung's rights are limited to the recognition of those Traditional Owner rights contained in s 9 of the *TOSA*. The *TOSA* is discussed in more detail in Chapter 7; however, it suffices to note here that water rights under the *TOSA* are also very limited.

At this stage, nothing suggests that other claimant groups will be able to negotiate for greater water rights than those contained in the Victorian *NTA* determinations and *TOSA* settlements made to date.

118 ALRC, *Connection to Country: Review of the Native Title Act 1993* (Cth), Report No 126 (2015).
119 Ibid ch 8.
120 Ibid 249 [8.105].
121 Ibid 248–249 [8.97]–[8.105].
122 This is likely to have been due to the scope of the report's terms of reference; see ibid 5–6.
123 Ibid 249 [8.105].
124 Commonwealth of Australia, 'Reforms to the *Native Title Act 1993* (Cth) Options paper, November 2017', 3.
125 The options do, however, consider providing for a more detailed register of ILUAs and for a register of subdivision P agreements, ibid 15–16.
126 There is nothing in the *TOSA* which precludes both a determination of native title and agreement being reached under the *TOSA*, the Gunai/Kurnai settlement being evidence of this.
127 As part of the *TOSA* settlement, the Dja Dja Wurrung withdrew their native title claim.

Concluding remarks

Despite the initial promise of both *Mabo* and the *NTA*, neither has lived up to expectations, particularly in relation to Indigenous participation in the management of water resources. Recognition of water rights under the *NTA* is limited to only those rights capable of being recognised by the common law, which is reflected in the limited scope of water rights recognised in determinations to date. The *NTA* provisions relating to procedural rights, as interpreted by the courts, further reduce the capacity of native title holders to participate in decisions regarding the management of water resources, to the point where participation is almost meaningless. Significant barriers for Indigenous groups, particularly in the more settled parts of Australia, to obtain recognition of any native title rights remain high. It is, therefore, apparent that as a legal mechanism for the recognition of Indigenous rights to participate in the management of water resources, the *NTA* is woefully inadequate.

Further, contrary to the sentiments contained in the preamble of the *NTA*, purporting fulfilment of Australia's international obligations to protect the rights of its Indigenous peoples, the *NTA* clearly falls well short of expectations, particularly regarding water rights.

Given the inadequacy of the *NTA*, are there other legal avenues by which Indigenous Australians can assert a right to participate in the management of water resources, and, if so, how effective might they be? These avenues form the subject of the next three chapters.

5 Other Commonwealth legislation relating to Indigenous participation in water management

Introduction

The *Native Title Act 1993* (Cth) ('*NTA*') was shown to be inadequate as a basis for facilitating Indigenous participation in water management, so we now turn our attention to other Commonwealth legislation that relates to or impacts on water management, namely the *Water Act 2007* (Cth) ('*Water Act 2007*') and the *Environment Protection and Biodiversity Conservation Act 1999* (Cth) ('*EPBC Act*').

Water Act 2007 (Cth)

Chapter 3 investigated Indigenous participation in the development of the *Water Act 2007*, concluding that until the recent amendments there had been minimal Indigenous participation.

A key management feature established by the *Water Act 2007* is the Murray–Darling Basin Authority ('MDBA') which prepares the Murray–Darling Basin Plan ('Basin Plan'), a catchment-wide planning framework for the sustainable management of water resources. In preparing the Basin Plan, the MDBA exercises its functions in a consultative manner,[1] which includes consultation with the Basin Community Committee ('BCC'), also established by the Act.[2] Another management feature established by the Act is the Commonwealth Environmental Water Holder, the function of which is to manage the Commonwealth environmental water holdings.[3]

The Basin states[4] continue to have a major role in water management through the preparation of water resource plans ('WRPs') which are required to give effect to the Basin Plan. State WRPs are accredited by the Minister.[5] A WRP can also

1 *Water Act 2007* ss 41–43.
2 Ibid s 202.
3 Ibid s 104.
4 Queensland, New South Wales, Victoria, South Australia and the Australian Capital Territory: ibid s 4.
5 The Minister can also adopt a WRP prepared by the Authority, if a Basin state has not given the Authority a WRP for accreditation, or the Minister has refused accreditation to a WRP or an amendment to a WRP: ibid s 68.

include other documents, insofar as they relate to the MDB water resources and the requirements of the Basin Plan.[6]

Those provisions of the *Water Act 2007* that relate to the preparation and content of the Basin Plan, and those which relate to the BCC are considered here.

Indigenous interests and the Basin Plan

Preparation of the Basin Plan

Section 21(4)(c) contains the relevant references, stating (in part) as follows:

> (4) Subject to subsections (1), (2) and (3), the Authority and the Minister must, in exercising their powers and performing their functions under this Division [ie preparing or amending a Basin Plan]: ...
> (c) have regard to the following:
> (i) the National Water Initiative;
> ...
> (v) social, cultural, Indigenous and other public benefit issues.

Subsections (i) and (v) are therefore relevant considerations to which the Authority (the MDBA) and the Minister must have regard in preparing or amending a Basin Plan. The reference to the *National Water Initiative* ('*NWI*') necessarily encompasses Indigenous interests, and as the deficiencies of the *NWI* in that regard have already been canvassed,[7] it suffices to say that they are very limited and do not form a core element of the *NWI*. Thus, the reference to the *NWI* in the *Water Act 2007* would appear to be tenuous insofar as it assists in the recognition of Indigenous interests in water management.

However, despite this, it is evident that the *NWI* did play a role in the recognition of Indigenous interests in the current *Basin Plan 2012* (Cth) ('*Basin Plan 2012*'). First, the *NWI* was confirmed by the MDBA at a general level as something to which it had regard,[8] and is referred to in schedule 1 of the *Basin Plan 2012*.[9] Second, those clauses of the *NWI* specifically relating to Indigenous interests are reflected to a substantial degree in the *Basin Plan 2012*, clearly indicating that the MDBA had regard to them.[10]

Indigenous issues are explicitly acknowledged in s 21(4)(c)(v) of the *Water Act 2007*. But they are merely one of a number of matters in sub-s (v) to which the MDBA and the Minister must have regard, the others being social, cultural and

6 MDBA, *Handbook for Practitioners: Water Resource Plan Requirements* (October 2013) 18–19.

7 See Chapter 3.

8 Letter from Craig Knowles, Chair, Murray–Darling Basin Authority to Tony Bourke, Minister for Sustainability, Environment, Water, Population and Communities, 21 November 2012, Attachment F.

9 *Basin Plan 2012* sch 1, [33], [40].

10 Clause 52(i) of the *NWI* is reflected in s 10.53(1)(c), cl 52(ii) is reflected in s 10.53 (1)(d), and cl 53 is reflected in s 10.53(1)(a).

other public benefit issues. This is compounded by the fact that sub-s (v) itself is only one of ten matters listed in s 21(4)(c). Given the administrative law principle that 'in the absence of any statutory indication of the weight to be given to various considerations, it is generally for the decision-maker and not the court to determine the appropriate weight to be given to each of the matters',[11] it is incumbent upon Indigenous groups to ensure that their issues are clearly articulated, so that they can be given the appropriate weight.

There are also other matters that must be taken into account, promoted or given effect to in the Basin Plan.[12] So Indigenous issues are but one of numerous matters to be considered in preparing the Basin Plan.

Further, although the MDBA and the Minister 'must ... have regard to' Indigenous issues, this is only a procedural requirement that will not necessarily result in a substantive outcome for Indigenous people. That is not to say that it is merely a 'tick-a-box' exercise: the MDBA and the Minister should give proper, genuine and realistic consideration to Indigenous issues.[13]

Despite these deficiencies, s 21 of the *Water Act 2007* is still valuable in that it explicitly refers to Indigenous issues as something to which the Minister and the MDBA must have regard when preparing or amending a Basin Plan. This puts those decision-makers on notice that Indigenous issues are clearly a relevant consideration and therefore cannot be ignored.

Content of the Basin Plan – section 22

Section 22(1)(b) requires the Basin Plan to provide information about Indigenous uses of Basin water resources. The benefit of this requirement is that those uses must be ascertained, which necessarily implies consultations with Indigenous people. However, there are no obligations regarding what is to be done with that information once it has been included in the Basin Plan. The fact that WRPs are required to be consistent with the Basin Plan does not extend to a requirement that WRPs be consistent with the uses to which the Basin water resources are put by Indigenous people.

Neither s 21 nor s 22 actively facilitates Indigenous participation in the management of the Basin's water resources, but they do provide some (albeit limited) scope for the recognition of Indigenous issues, and of Indigenous uses of water.

In that respect, for the MDBA and the Minister to 'have regard to' Indigenous issues and for the Basin Plan to include information about Indigenous uses of water resources, there must inevitably be input from Indigenous people. Accordingly, despite the limited nature of these provisions, there were substantial consultations with Indigenous groups in the preparation of the *Basin Plan 2012*.[14]

11 *Minister for Aboriginal Affairs v Peko-Wallsend Ltd* (1986) 162 CLR 24, 41 (Mason J).
12 *Water Act 2007* s 21(4)(a), (b), s 21(1)–(3).
13 *Hindi v Minister for Immigration and Ethnic Affairs* (1988) 20 FCR 1.
14 MDBA, *A Yarn on the River Getting Aboriginal Voices into the Basin Plan* (2011).

As a result, the *Basin Plan 2012* contains in the front matter an acknowledgement of the Traditional Owners of the MDB. Chapter 10 of the *Basin Plan 2012* relates to the requirements for WRPs, outlined earlier. It contains part 14 – Indigenous values and uses, consisting of s 10.52–55. This part requires that WRPs must identify the objectives of and the outcomes desired by Indigenous peoples in relation to the management of water resources, and that regard must be had to Indigenous values and uses in determining them.

Section 10.53 relates to consultation and requires that '[a] water resource plan must be prepared having regard to the views of relevant Indigenous organisations with respect to the matters identified under section 10.52 and the following matters'. Those matters include native title, Aboriginal heritage, Indigenous social, cultural, spiritual and customary objectives, Indigenous representation, Indigenous participation, and risks to Indigenous values and uses (but notably not economic or commercial uses). A WRP must also be prepared having regard to the views of Indigenous people with respect to cultural flows.[15]

As mentioned earlier in relation to the preparation of the Basin Plan, the reference to 'having regard to' is a procedural requirement and does not mandate a particular outcome. This is noted in the *Basin Plan 2012.* [16] And although WRPs must identify the management objectives and outcomes desired by Indigenous people, there is no mandatory requirement that steps be put in place to achieve those outcomes.[17]

The Basin Plan is a legislative instrument,[18] and therefore has legislative effect. The *Water Act 2007* provides that all WRPs must be consistent with the Basin Plan.[19] WRP compliance with the *Basin Plan 2012* is to occur by 1 July 2019.[20] Therefore, Victoria's WRPs under the *Water Act 1989* (Vic) that relate to the MDB are required to comply with these provisions.[21] Accordingly, the *Water Act 2007* and the *Basin Plan 2012* should compel Victoria to consider Indigenous issues, insofar as they fall within the MDB. One would hope that this would have a consequential effect on the rest of Victoria.

Accreditation of a WRP requires the Minster to have regard to the Basin Plan and the extent to which the WRP is consistent with the Basin Plan.[22] This suggests that the Minister may refuse to accredit a WRP if it is not consistent with part 14 of chapter 10. However, the *Water Act 2007* also provides that '[i]n determining whether the water resource plan is consistent with the relevant Basin

15 *Basin Plan 2012* s 10.54.
16 Ibid note to s 1.7.
17 Ibid s 10.52(3).
18 *Water Act 2007* s 33.
19 Ibid s 55(2).
20 MDBA, *Handbook for Practitioners*, above n 6, 2.
21 There is no specific document in Victoria's *Water Act 1989* called a WRP. Instead, WRPs comprise various documents, including Sustainable Water Strategies, Regional Water Strategies, Seasonal Watering Plans, and bulk entitlements. These kinds of arrangements are acknowledged by the MDBA: ibid 3.
22 *Water Act 2007* s 56(1).

Plan, regard must be had to the legislative framework within which the water resource plan operates'.[23] Proposed amendments to Victoria's water management legislation[24] suggests that Victoria's WRPs will likely address the matters set out in part 14 of chapter 10 of the *Basin Plan 2012*. In addition, Victoria's current Sustainable Water Strategies and Regional Water Strategies refer to Indigenous interests.[25] As they would be considered to form part of Victoria's WRPs, Victoria is already complying with the *Basin Plan 2012* to some degree.

Basin Community Committee

There is scope for more direct Indigenous participation in water management via the BCC, established pursuant to s 202. The BCC is an advisory committee, the function of which is:

> to advise the Authority about the performance of the Authority's functions, including advising about:
> (a) engaging the community in the preparation of each draft Basin Plan; and
> (b) community matters relating to the Basin water resources; and
> (c) matters referred to the Committee by the Authority.[26]

In addition, the MDBA must consult with the BCC in preparing the Basin Plan,[27] in preparing any amendment of the Basin Plan,[28] or in preparing a discussion paper for any review of the Basin Plan.[29]

Membership of the BCC must include 'at least 2 Indigenous persons with expertise in Indigenous matters relevant to the Basin's water resources'.[30] It must also include a member of the MDBA, and eight individuals who are water users.[31] The BCC can have up to 17 members, including the Chair.

To be eligible for appointment to the BCC, an individual 'must have a high level of expertise or interest in: (a) community, indigenous or local government matters relevant to the Basin's water resources'.[32] The BCC must also establish 'an Indigenous water subcommittee, to guide the consideration of Indigenous matters relevant to the Basin's water resources'.[33]

As noted, Indigenous experts are only two of up to 17 members of the BCC, and at least eight of those 17 members must be water users, as defined in s 204

23 Ibid s 55(3).
24 Water and Catchment Legislation Amendment Bill 2017 (Vic).
25 See Chapter 6.
26 *Water Act 2007* s 202(2).
27 Ibid s 43(1)(c).
28 Ibid s 46(1)(c).
29 Ibid s 51(2)(c).
30 Ibid s 202(5)(c).
31 Ibid s 204(7).
32 Ibid s 204(3)(a).
33 Ibid s 202(3)(c).

(7). That definition does not include Indigenous water users. Therefore, any advice provided to the MDBA is likely to be weighed towards non-Indigenous community interests. That is not to say that non-Indigenous members of the BCC will exclusively focus on their own interests and that any advice provided to the MDBA will not include advice on Indigenous matters, but given the wide range of community matters that are likely to arise in relation to the MDB, the probability of Indigenous matters being given sufficient weight is lessened.

The BCC also acts only in an advisory capacity; the MDBA must consult with but is not required to follow the advice of the BCC. So, in terms of participation in water management, the BCC is not a decision-maker; that role still resides with the MDBA. Accordingly, it would clearly be of value to have Indigenous representation on the MDBA. In that regard, eligibility for membership of the MDBA is based on having a high level of expertise in a field relevant to the MDBA's functions, which includes 'Indigenous matters relevant to Basin water resources'.[34] It does not require that the person with that expertise be Indigenous, but in practice this would likely be the case.

Thus, as far as the *Water Act 2007* is concerned, there are limited legislative requirements ensuring Indigenous participation in water management, those requirements being confined to participation via the BCC whose membership is heavily weighted towards non-Indigenous interests, and whose advice the MDBA is not obliged to follow, and potential for a person with expertise in Indigenous matters relevant to Basin water resources to be on the MDBA. The *Water Act 2007* does, however, explicitly require the consideration of Indigenous interests in the preparation and amendment of the Basin Plan, and for the content of the Plan to include information about Indigenous uses of Basin water resources, which necessarily involves consultation with Indigenous people.

Environment Protection and Biodiversity Conservation Act 1999 (Cth)

As with the regulation and management of water, the Commonwealth does not have specific powers in the *Constitution* over the environment.[35] However, the *EPBC Act* is clearly based upon the external affairs power[36] and the corporations power.[37]

34 Ibid s 178(3)(g).

35 See, eg, Senate Environment, Communications, Information Technology and the Arts References Committee, Parliament of Australia, *Commonwealth Environment Powers* (1999) 7 [2.14] which notes numerous heads of power relied on by the Commonwealth to enact environmental legislation. For more information on the constitutional basis for the *EPBC Act*, see Catherine Branson, 'The *Environmental Protection and Biodiversity Conservation Act 1999* – Some Key Constitutional and Administrative Issues' (1999) 6(1) *Australasian Journal of Natural Resources Law and Policy* 33.

36 *Commonwealth Constitution*, s 51(xxix). See reference in the *EPBC Act* s 3(1)(e) to Australia's international environmental responsibilities, and express references to various international environmental conventions to which Australia is a party.

37 *Commonwealth Constitution*, s 51(xx). There are numerous references in the *EPBC Act* to constitutional corporations, defined in s 528 of the *EPBC Act* as 'a corporation to which paragraph 51(xx) of the *Constitution* applies'.

The objects of the *EPBC Act* are contained in s 3(1), with three of the eight objects referring to Indigenous people.[38] The means by which the *EPBC Act* achieves these objects include by 'recognising and promoting indigenous peoples' role in, and knowledge of, the conservation and ecologically sustainable use of biodiversity'.[39] So, at the outset, the *EPBC Act* clearly identifies Indigenous people as having a role in the protection of the environment and biodiversity conservation that should be recognised and promoted.

Indigenous representation on advisory committees

This is further evidenced by the establishment of an Indigenous Advisory Committee ('IAC'), the function of which 'is to advise the Minister on the operation of the Act, taking into account the significance of indigenous peoples' knowledge of the management of land and the conservation and sustainable use of biodiversity'.[40]

The IAC, however, is

> an advisory and not a decision-making body and it is not a consultative mechanism to seek the views of stakeholders. It does not advocate on behalf of Indigenous communities, stakeholders or funded proponents. It provides strategic advice at the request of the Minister.[41]

The IAC advises on Indigenous issues relating to the whole of the *EPBC Act*, which is not limited to water management. Under the IAC's terms of reference, its role also includes advising on other matters that affect Aboriginal and Torres Strait Islander people in the environment portfolio and reporting on matters of interest to the Department.[42] Given that its members are appointed on a part-time basis[43] and meet approximately twice a year (although there are regular teleconferences), the IAC is limited in both time and resources in its ability to provide comprehensive strategic advice on the many Indigenous issues arising under both the *EPBC Act* and within the Department generally.[44]

The IAC is one of three advisory committees established under the *EPBC Act*, the others being the Threatened Species Scientific Committee ('TSSC') and the Independent Expert Scientific Committee on Coal Seam Gas and Large Coal

38 *EPBC Act* s 3(1)(d), (f), (g).
39 Ibid s 3(2)(g)(iii).
40 Ibid s 505B(1).
41 Department of the Environment and Energy (Cth), Indigenous Environment Committee <www.environment.gov.au/epbc/advisory-committees/iac>.
42 Ibid.
43 *EPBC Act* s 505A(3).
44 See, eg, Indigenous Advisory Committee, *Indigenous Advisory Committee Meeting 25 Briefing, 23–25 October 2012 Cairns, Queensland* (2013), Department of the Environment and Energy (Cth), *Indigenous Advisory Committee* <www.environment.gov.au/epbc/advisory-committees/iac>.

Mining Development. There is no requirement for Indigenous representation on either of these committees. For many years there was a Biological Diversity Advisory Committee ('BDAC'), the membership of which had to include a representative of Indigenous peoples, but this committee was abolished in 2016.[45]

The Australian Heritage Council ('AHC'), established pursuant to the *Australian Heritage Council Act 2003* (Cth) ('*AHC Act*') also has an advisory role under the *EPBC Act*, in relation to National Heritage places (discussed further below). The Minister, in appointing members to the AHC 'must ensure that: ... 2 of them are indigenous persons with substantial experience or expertise concerning indigenous heritage, at least one of whom represents the interests of indigenous people'.[46] However the ultimate responsibility for decision-making rests with the Minister, although it can be (and usually is) delegated.[47]

So there is clearly a role for Indigenous people in the operation of the *EPBC Act* via the IAC and the AHC. However, the role of these committees is advisory only; the Minister is not obliged to follow their advice or implement their recommendations. These committees also have practical limitations which inhibit them from being an effective mechanism for ensuring that the interests of Indigenous people in water management are appropriately considered. However, there is some scope for Indigenous participation in specific procedures under the *EPBC Act*, to which we now turn.

Indigenous participation in water management under the EPBC Act

The focus of the *EPBC Act* is on matters of National Environmental Significance ('MNES').[48] There are nine specified MNES.[49] Two of them, namely matters relating to World Heritage or National Heritage areas, may include areas on which water sources/resources are located. Wetlands of international importance (Ramsar sites) are clearly water related, as is the protection of water resources from coal seam gas and large coal mining development.

These areas may also be of importance to Indigenous people. Kakadu National Park in the Northern Territory is a prime example of a World Heritage area of significance to Indigenous people, containing numerous water sources. In Victoria, the Gippsland Lakes are a declared Ramsar site, and are within the recognised traditional country of the Gunaikurnai people.[50]

In relation to National Heritage sites, there are several places in Victoria on the National Heritage List of importance to Indigenous people, of which at least one, the Budj Bim National Heritage Landscape at Lake Condah, is a significant water

45 *Omnibus Repeal Day (Autumn 2015) Act 2016* (Cth) sch 2, pt 2.
46 *AHC Act* s 7(3)(c).
47 *EPBC Act* s 515.
48 Ibid s 3(a).
49 Ibid ch 2 pt 3 div 1.
50 As native title holders pursuant to a native title determination under the *NTA* and as the Traditional Owners pursuant to a Recognition and Settlement Agreement made under the *TOSA*.

site.[51] Budj Bim is in the traditional country of the Gunditjmara people, who were recognised as the native title holders in 2007. There are currently no relevant World Heritage places in Victoria; however, Budj Bim has been accepted for World Heritage assessment, a step towards it being formally recommended for World Heritage status.[52]

If an action is likely to have a significant impact[53] on one or more MNES, then generally an environmental approval must be obtained, either from the Commonwealth Environment Minister, or via an accredited state environmental approval process where there is a bilateral agreement in place[54] (of which there were none as at May 2018). Actions that are not likely to have a significant impact are subject only to relevant state laws, and the *NTA*. [55] As to whether there is any specific role for Indigenous people, in summary, where an approval is required under the *EPBC Act* for actions which impact upon those MNES which involve (or could potentially involve) water resources, there are only public consultation processes by which Indigenous people can provide their views. These are inadequate because they do not consider the specific needs of Indigenous people, or their status as First Peoples.

Prior to deciding whether or not to approve an action, the Minister is required to assess the environmental impact of the proposed action. This can also be delegated to the states via bilateral agreements. Each of the states and territories has a bilateral agreement relating to environmental impact assessments.[56]

In Victoria, the accredited assessment processes in its bilateral agreement include the procedures contained in the *Environment Effects Act 1978* (Vic) ('*EE Act*') for the preparation of an Environmental Effects Statement ('EES'). The *EE Act* contains no requirement to consult with Indigenous people in the preparation of an EES; however, Ministerial Guidelines made under the *EE Act* encourage Indigenous consultation.[57] Further, under the *Aboriginal Heritage Act 2006* (Vic), a cultural heritage management plan must be prepared for any project

51 Department of the Environment, Water, Heritage and the Arts, *Australia's National Heritage* (2010) 38, 76 <www.environment.gov.au/heritage/publications/australias-national-heritage-2010>.
52 Josh Frydenberg, 'Historic Step Towards Global Recognition for Budj Bim' (Media release, 2 March 2018).
53 'Significant impact' is not defined in the *EPBC Act*; however, the Commonwealth has produced administrative guidelines: Department of the Environment, Water, Heritage and the Arts, *Matters of National Environmental Significance, Significant Impact Guidelines 1.1, Environment Protection and Biodiversity Conservation Act 1999* (2009).
54 Bilateral agreements are provided for in the *EPBC Act* pt 5.
55 Ibid s 8(2)(b).
56 Department of the Environment and Energy (Cth), *One Stop Shop for Environmental Approvals* <www.environment.gov.au/topics/about-us/legislation/environment-protection-and-biodiversity-conservation-act-1999/one-stop>.
57 *EE Act* s 10. Department of Sustainability and Environment (Vic), *Ministerial Guidelines for Assessment of Environmental Effects under the Environment Effects Act 1978* (7th edn, June 2006) 13.

which requires an EES.[58] So an EES involving water will require consultation with Indigenous people.[59]

The bilateral agreement also accredits the approvals process for applications made under ss 36, 51 and 67 of the *Water Act 1989* (Vic),[60] which can be used where an EES is not required.[61] None of these provisions require consultation with Indigenous people, only public consultation processes which, like the Commonwealth approvals processes, are inadequate because they do not consider the specific needs of Indigenous people or their status as First Peoples.

National Heritage places

National Heritage places are an MNES and, as noted earlier, there are several Indigenous places in Victoria on the National Heritage List. There are various requirements for Indigenous participation in relation to both the listing and management of National Heritage places.[62]

Listing of National Heritage places

Nominations received by the Minister are forwarded to the AHC which prepares a priority list of eligible sites to be assessed for inclusion on the National Heritage List.[63] The list is then finalised by the Minister and published by the AHC.[64] The AHC must then seek comments on the finalised priority list, and conduct an assessment of whether the place meets the National Heritage criteria, which includes 'indigenous heritage values'.[65] If the AHC considers that a place might have Indigenous heritage values, then it must identify any Indigenous person who may have rights or interests in the place (or body representing such persons) and give them 20 days to comment on whether the place should be included on the list.[66]

So Indigenous participation in the listing of National Heritage places is via three avenues: first by nominating a place for listing, second by way of the AHC (which has two Indigenous members)[67] and third, by having been identified by the AHC as having rights and interests in the nominated place. For Indigenous people, it is the first and third avenues of the nomination process that enables them to have a

58 *Aboriginal Heritage Act 2006* (Vic) s 49.
59 See also Chapter 7.
60 Applications by a water authority for bulk entitlements (s 36), applications for a licence to take and use (s 51) and a licence to construct works (s 67).
61 Department of Environment, Land, Water and Planning (Vic), *Environment Assessment* <www.planning.vic.gov.au/environment-assessment/environmental-assessment-bilateral-agreement>.
62 *EPBC Act* ch 5 pt 15 div 1A.
63 Ibid ss 324J, 324JA, 324JB.
64 Ibid ss 324JE, 324JF.
65 Ibid s 324D(3)(b).
66 Ibid ss 324JH(5), (7).
67 *AHC Act* s 7(3)(c).

say over their land and waters, as the AHC's Indigenous members do not represent specific Indigenous groups. However, because the Minister has discretion whether to include a place on the list,[68] there are limits to the nomination process for Indigenous people, as there is no guarantee that a place that has been nominated and subsequently recommended for inclusion on the list by the AHC will be approved.

Management of National Heritage places, Ramsar sites and World Heritage sites

In relation to National Heritage places on Commonwealth areas, the Minister is required to prepare a management plan,[69] and must seek comments from the relevant Indigenous people in relation to its draft.[70] For National Heritage places that are not entirely within a Commonwealth area, the Commonwealth must try to ensure that a plan for managing the place is prepared and implemented in cooperation with the state or territory, consistent with the National Heritage management principles ('NHMPs').[71] National Heritage places that do not fall within a Commonwealth area are not subject to such statutory requirements, but their management would still be guided by the NHMPs.

Principle 6 is relevant to Indigenous participation, stating:

> Indigenous people are the primary source of information on the value of their heritage and the active participation of indigenous people in identification, assessment and management is integral to the effective protection of indigenous heritage values.

As there are limited Commonwealth areas in Victoria (and therefore the management plan provisions have little application), only the NHMPs (and principle 6 in particular) are relevant to Indigenous participation in the management of places on the National Heritage List. Where the listing of a place as a National Heritage place was initiated by a Traditional Owner group, such as the Budj Bim National Heritage Landscape by the Gunditjmara people, they are necessarily going to be involved in its management.[72]

The management principles for Ramsar sites,[73] however, do not contain any requirement for Indigenous participation, nor do they contain any equivalent of principle 6. It is a similar situation in relation to World Heritage sites.[74] Therefore,

68 *EPBC Act* s 324JJ.

69 Ibid s 324S.

70 *EPBC Regulations 2000* (Cth) reg 10.01D(c)(ii) ('*EPBC Regulations*').

71 *EPBC Act* s 324X(2). NHMPs are contained in the *EPBC Regulations* sch 5B.

72 The Budj Bim National Heritage Landscape is managed on behalf of the Gunditjmara by the Gundtij Mirring Traditional Owners Aboriginal Corporation. Parts of it are owned in freehold by the Gunditjmara and are managed as Indigenous Protected Areas.

73 *EPBC Regulations* sch 6.

74 Ibid reg 10.01, sch 5. It is likely that World Heritage sites are also National Heritage places, so in practice this is not likely to be an issue. For example, Kakadu National Park is both a National Heritage place and World Heritage site: Department of the

the *EPBC Act* provides a greater legislative basis for Indigenous participation in the management of National Heritage places than of Ramsar and World Heritage sites.

Reform of the EPBC Act

The *EPBC Act* was the subject of an independent statutory review[75] commencing in 2008.[76] The final report[77] contained 71 recommendations of which five were of relevance to Indigenous people. In its response to the final report,[78] the Commonwealth government agreed to four of the five recommendations, and in relation to one recommendation was equivocal.

One of the agreed recommendations was to extend the requirement for informed consent on the use of Indigenous knowledge to non-commercial purposes (currently only commercial use is covered);[79] another was to disband the BDAC and transfer its functions to the TSSC.[80] The third recommendation was to enable better coordination of the *EPBC Act* with legislation relating to Indigenous interests,[81] and the fourth sought better coordination of and information exchange between relevant statutory bodies.[82]

The recommendation about which the government was equivocal involved improving public participation by (among other things) strengthening involvement of Indigenous peoples in the working of the Act.[83]

The *EPBC Act* has been amended since, most notably to include the protection of water resources from coal seam gas and large coal mining development as an additional MNES. However, the only recommendation relating to Indigenous interests that has been implemented is to disband the BDAC.[84]

Concluding remarks

The Commonwealth's *Water Act 2007* and *EPBC Act* are somewhat uncoordinated when it comes to the legislative recognition of Indigenous interests in water

Environment and Energy, *About Australia's Heritage* <www.environment.gov.au/heritage/about>.

75 *EPBC Act* s 522A.

76 Peter Garrett, 'EPBC Act Under Independent Review' (Media Release, 31 October 2008).

77 Allan Hawke, *The Australian Environment Act: Report of the Independent Review of the Environment Protection and Biodiversity Conservation Act 1999* (Department of the Environment, Water, Heritage and the Arts, October 2009).

78 *Australian Government Response to the Report of the Independent Review of the Environment Protection and Biodiversity Conservation Act 1999* (Department of Sustainability, Environment, Water, Population and Communities, 2011).

79 Ibid recommendation 22(4).

80 Ibid recommendation 68.

81 Ibid recommendation 64.

82 Ibid recommendation 69.

83 Ibid recommendation 45(2)(b).

84 *Omnibus Repeal Day (Autumn 2015) Act 2016*, sch 2 pt 2.

management. However, there are some specific requirements for Indigenous issues to be considered by decision-makers. The *Water Act 2007* requires that decision-makers under the Act have regard to Indigenous issues in the preparation of the Basin Plan and that the Basin Plan provide details about Indigenous uses of Basin water resources. It also requires WRPs prepared by the states to be consistent with the Basin Plan, thus compelling the states to consider the Indigenous elements of the Basin Plan (albeit within the context of the state's legislative framework).

Representation is also mandated, the Commonwealth legislation requiring that membership of various advisory committees established under each Act contain people with experience in Indigenous issues. Under the *EPBC Act*, all the IAC's members are Indigenous. The BCC under the *Water Act 2007* must include at least two Indigenous people with expertise in Indigenous matters relevant to Basin water resources.

However, given that their respective spheres of operation are constrained because of the limitation on Commonwealth powers under the *Constitution*, the *Water Act 2007* and the *EPBC Act* are necessarily restricted in what they can provide by way of Indigenous participation in water management. To improve Indigenous participation in any meaningful way, we must turn to state legislation, under which the bulk of the substantive decision-making occurs.

Part D

Case study

Victoria

6 Indigenous participation in water management via the *Water Act 1989* (Vic)

Introduction

The *Water Act 1989* (Vic) ('*Water Act 1989*') is the primary legislative instrument governing water access, use and management in Victoria. Accordingly, for Victoria's Traditional Owners to participate in the management of water resources it is essential that their interests be recognised in that Act.

This chapter will analyse the *Water Act 1989*, first outlining the background and policy context, then, in relation to provisions which explicitly recognise Traditional Owner water rights, their adequacy in facilitating Traditional Owners to participate in the management of water resources, and next in relation to how other provisions of the Act might be (or have been) used to facilitate Traditional Owner participation in water management. It will also touch on the *Catchment and Land Protection Act 1994* ('*CALP Act*') given the linkages between the two Acts in relation to water management.

Background and policy context

As noted in Chapter 3, the Water *Act 1989* was intended to replace an outdated, complicated and often inconsistent regulatory regime by consolidating numerous Acts into one.

The development of the original *Water Act 1989* was also the product of significant public consultation, which occurred without Traditional Owners being identified as stakeholders whose interests in water management needed to be considered. This resulted in no recognition of Traditional Owner interests in the *Water Act 1989*, and a bias in favour of those with existing water allocations, a bias which, despite recent amendments, is still reflected in the purposes of the Act.[1] This in effect amounts to a form of structural dispossession, whereby due to the dispossession of Traditional Owners of their land following the acquisition of sovereignty, they were not part of any initial water allocations, as those allocations were linked to land ownership and riparian rights.[2] With the decoupling of water

1 *Water Act 1989* s 1(m).
2 In relation to NSW's water legislation: see Tony McAvoy, 'Water – Fluid Perceptions' (2006) 1(2) *Transforming Cultures eJournal* 97, 101.

rights from land ownership, any subsequent recognition of Traditional Owner rights in land would not include water rights. This is further compounded by the limited scope of land rights in Victoria (see Chapter 7), and is apparent in the recognition of native title rights in water, as they are subordinate to those water rights existing at the time of the determination (as are all native title rights). These comments relate generally to the allocation of water resources, but have a consequential effect on the management of water resources in that those with allocation rights are more likely to be given the opportunity to participate in decisions regarding water management. As we shall see, this is reflected in the *Water Act 1989*.

Following the enactment of the *Water Act 1989*, there were a number of policy developments in the national sphere. As outlined in Chapter 3, these developments led to the *National Water Initiative* ('*NWI*'),[3] in which limited recognition of Indigenous interests has occurred. Implementation of the *NWI* requires states to take steps to provide for Indigenous access to water resources by, among other things, ensuring the inclusion of Indigenous representation in water planning wherever possible and by taking into account native title rights and interests in water.[4] In addition, one of the outcomes of the water access entitlements and planning framework element of the *NWI* is that the parties to the *NWI* will 'recognise indigenous needs in relation to water access and management'.[5]

In Victoria, steps to implement these objectives and achieve this outcome included at first instance the preparation of various strategies relating to water management, all of which specifically refer to or have sections relating to Indigenous water issues.[6] Some Catchment Management Authorities ('CMAs') and government departments also established Indigenous reference or advisory groups to provide input into their decision-making processes.[7] However, legislative amendment recognising Traditional Owner water rights has been much slower to occur.

3 *Intergovernmental Agreement on a National Water Initiative* (25 June 2004).
4 *NWI* cls 52–54.
5 Ibid cl (ix).
6 These include Department of Sustainability and Environment ('DSE'), *Gippsland Region Sustainable Water Strategy* (November 2011) ('*GRSWS*') s 4.3; DSE, *Northern Region Sustainable Water Strategy* (November 2009) s 9.1.1; DSE, *Sustainable Water Strategy Central Region – Action to 2055* (October 2006) ch 2, 20; DSE, *Western Region Sustainable Water Strategy* (November 2011) s 3.5: Department of Environment, Land, Water and Planning ('DELWP'), *Sustainable Water Strategies* <www.water.vic.gov.au/planning-and-entitlements/sustainable-water-strategies>; and the Department of Environment and Primary Industries, *Improving Our Waterways — Victorian Waterway Management Strategy* (September 2013) ('*Victorian Waterway Management Strategy*') ch 6; DELWP, *Strategy and Planning* <www.water.vic.gov.au/waterways-and-catchments/rivers-estuaries-and-waterways/strategies-and-planning>.
7 See, eg, *GRSWS*, above n 6, 85. See also National Water Commission, *Australia's Water Blueprint: National Reform Assessment 2014*, 233.

Traditional Owner rights under the *Water Act 1989* (Vic)

It was not until 2010 that the *Water Act 1989* was amended to explicitly recognise Traditional Owner rights in water. This amendment occurred due to the enactment of the *Traditional Owner Settlement Act 2010* (Vic) ('*TOSA*') the background to which is detailed in Chapter 7.

The *TOSA* inserted s 8A into the *Water Act 1989*. That section currently reads as follows:

> **8A Traditional owner agreement for natural resources**
> If a traditional owner group entity has an agreement under Part 6 of the **Traditional Owner Settlement Act 2010**, a person who is a member of a traditional owner group bound by the agreement has the right to take and use water on the land that is subject to the agreement—
> (a) in accordance with the agreement; and
> (b) if the water is to be taken from a place from which water may be taken under section 8(1).

There are a couple of points to note about this provision. The first is that it says nothing about Traditional Owner rights to be involved in the *management* of water resources; it merely gives Traditional Owners the right to *take and use* water. Second, the purposes for which a Traditional Owner can take and use water are confined to traditional purposes, unlike some other resources.[8] And, finally, it refers only to Traditional Owner groups with an agreement under the *TOSA*; it does not mention native title holders under the *NTA*.

Furthermore, given that s 8(1) of the *Water Act 1989* allows any person (including Traditional Owners) to take water from a bore or waterway to which they have access for domestic and stock purposes, s 8A would appear to add very little to rights which Traditional Owners already have under s 8(1). Indeed, the *Gippsland Region Sustainable Water Strategy* explicitly acknowledges this similarity.[9]

Thus, s 8A is largely symbolic. Although we should not underestimate the importance of symbolism, it needs to be accompanied by practical recognition. The provision on its own has no effect on the operation of the *Water Act 1989*; it creates no obligations on water authorities or other water users to consider Indigenous water issues. Nor does it give Traditional Owners a voice in water management.

However, if the Water and Catchment Legislation Amendment Bill 2017 (Vic) ('WCL Amendment Bill') becomes law, the position in relation to Indigenous participation in water management will be substantially improved.

8 Amendments to the *TOSA* in 2013 now allow the use of flora and forest produce for commercial purposes, but not water resources.

9 *GRSWS*, above n 6, 82.

Consultative committees

The *Water Act 1989* establishes a number of consultative committees to assist in the preparation of various documents that guide water management in Victoria. Currently, there is no requirement for any of the committees to have an Indigenous member. The proposed amendments will require all but one of the committees to include at least one Indigenous representative.[10]

The documents at the highest level are Sustainable Water Strategies ('SWSs') which 'must provide for the strategic planning of the use of water resources in the region to which it applies'.[11] SWSs were introduced into the *Water Act 1989* in 2005 to implement both the water reforms outlined in the government's 2004 *White Paper* and the objectives of the *NWI*. [12] It was clearly the government's view at the time that Indigenous groups would participate in the development of SWSs,[13] but this was not legislatively mandated.

The four SWSs currently in existence cover all of Victoria;[14] however, there were no Traditional Owners represented on any of the consultative committees which developed them.[15] Nonetheless, each of them refers to Indigenous, Aboriginal or Traditional Owner interests.[16]

The reference to Indigenous, Aboriginal or Traditional Owner interests in the four SWSs, despite no Indigenous representatives on the consultative committees, may be attributed to a number of factors, the first being the existence of the Department of Sustainability and Environment's ('DSE') (as it was then called) [17]*Indigenous Partnership Framework* ('*IPF*'), which notes that '[a] key aim of the *IPF* is to facilitate the full, effective and equitable participation of Traditional Owners and Indigenous people in all aspects of land and natural resources management'.[18] Second, the existence of native title claims (and subsequent determinations in some areas) over much of the state for many years meant that Traditional Owners were asserting their right to speak for country throughout the period of preparation of these SWSs. DSE, as well as being represented on all the consultative committees, was heavily involved in native title matters.[19] Third, and

10 The exception will be Water Management Scheme Committees, established pursuant to s 214 of the *Water Act 1989*.
11 *Water Act 1989* s 22C(1).
12 Victoria, *Parliamentary Debates*, Legislative Assembly, 6 October 2005, 1335 (John Pandazopoulos).
13 Ibid 1336.
14 The state was divided into four regions for this purpose: Gippsland, Western, Northern and Central.
15 There was no consultative committee appointed for the Central region SWS.
16 See above n 6.
17 The relevant department is currently called the Department of Environment, Land, Water and Planning.
18 DSE, *Indigenous Partnership Framework 2007–10* (2007) 1. This Framework has since been replaced by the following plan: Department of Environment and Primary Industries, *Meerreeng Wanga: Aboriginal Inclusion Plan 2014–2019* (2014).
19 DSE was the government department which dealt with future act notifications under the *NTA*. It was also (along with the Department of Justice) heavily involved in

relatedly, the development of the Victorian Native Title Settlement Framework ('VNTSF'), which resulted in the *TOSA*, clearly played a role as well. DSE was also a major participant in the development of the VNTSF.[20] Finally, the three SWSs produced after the enactment of the *Aboriginal Heritage Act 2006* (Vic) ('*AHA*')[21] refer to Registered Aboriginal Parties ('RAPs') and RAP applicants as being among the Traditional Owner groups consulted in the development of the SWSs. This suggests that the *AHA* was also a factor in ensuring that Traditional Owners were consulted.

Some of the participants on the consultative committees (particularly the DSE representatives) would have had contact with Traditional Owner groups through one or more of these processes and would thus have been aware of the need to consult with Traditional Owners.

Accordingly, although at the time there was no legal requirement in the *Water Act 1989* to consult with Traditional Owners, the existence of other Acts, such as the *NTA*, the *TOSA* and the *AHA*, which recognise Traditional Owners as having various rights over land and waters, and which themselves require some form of Traditional Owner engagement, was likely to have been a motivating factor behind the SWS consultations with Traditional Owners.

The proposed amendments will require the Minister to ensure that as far as possible, the membership of an SWS consultative committee includes at least one Aboriginal person residing in Victoria who has knowledge or experience in the matters to be covered in a Sustainable Water Strategy.[22] The content of an SWS will also be required to 'consider opportunities to provide for … Aboriginal cultural values and uses of waterways in the region for which the Strategy applies',[23] thus providing further recognition of Indigenous interests in water.

Finally, the draft strategies produced by SWS consultative committees are also subject to a public consultation process whereby they are notified in local newspapers and comments are invited from the general public by a specific date.[24] The proposed amendments will require specified Aboriginal parties to be separately

settlement negotiations. DSE also published the draft and final strategies, and was involved in commissioning the technical reports on Indigenous engagement, and therefore it would appear that it also acted in a secretarial capacity for the preparation of the draft strategies.

20 Steering Committee for the Development of a Native Title Settlement Framework, *Report of the Steering Committee for the Development of a Native Title Settlement Framework* (December 2008) 1. This is also evidenced by the fact that the draft SWSs for the Western and Gippsland regions both refer to the VNTSF: *Draft Western Region Sustainable Water Strategy for Community Comment* (DSE, March 2010) 21; *Draft Gippsland Region Sustainable Water Strategy for Community Comment* (DSE, August 2010) 7.

21 The *Central Region SWS* was published in 2006, the same year as the enactment of the *AHA*.

22 WCL Amendment Bill cl 8.

23 Ibid cl 7.

24 *Water Act 1989* s 22E.

notified.[25] Thus, participation by Traditional Owners could occur in this context as well.

But given that there are no more regions for which an SWS needs to be developed, the focus of Traditional Owners should turn to participation in any SWS review, which will be made much easier if the proposed amendments become law. The review procedure will the same as that for the preparation of the initial SWS (with any necessary modifications),[26] with the addition of mandated Aboriginal representation on SWS review consultative committees, a distinct improvement on the original committees.

This analysis shows that despite the lack of representation on the original SWS consultative committees, there was some recognition of Traditional Owner interests in water management. Other Acts are likely to have influenced to some degree the level of Traditional Owner consultation, at least in relation to the preparation of SWSs, thus ensuring that Traditional Owner interests, if not at the forefront, at least were not neglected. The value of Traditional Owner involvement has now been clearly recognised by the proposed amendments to explicitly require the inclusion of Traditional Owners on SWS committees. This proposed strengthening of the role of Traditional Owners in the *Water Act 1989* will further enhance their participation in water management, enabling the prioritisation of Traditional Owner interests beyond that as members of the public or just another stakeholder. This may help to ameliorate some of the structural dispossession that has been a feature of Victoria's water management regime to date.

The remaining three consultative committees established by the *Water Act 1989* relate to water supply protection areas ('WSPAs'), long term water resource assessments ('LTWRAs') and water management schemes ('WMSs').

The WSPA provisions[27] were designed 'to ensure equitable distribution of water supplies in overcommitted catchments', allowing the Minister to declare a WSPA 'where overcommitment threatens the continued availability of water'.[28]

The Minister may declare an area to be a WSPA for the protection of groundwater or surface water resources in the area, or both.[29] If a declaration is made, the Minister must appoint a consultative committee to prepare a draft management plan for the area.[30] The object of such a management plan is 'to make sure that the water resources of the relevant water supply protection area are managed in an equitable manner and so as to ensure the long term sustainability of those resources'.[31] In preparing a draft management plan, a consultative committee 'must take into account any draft or approved management plan that applies to the area or part of the area'.[32] Although not specified, this would include a joint

25 WCL Amendment Bill cl 9.
26 *Water Act 1989* s 22I(4).
27 Ibid pt 3 div 3.
28 Victoria, *Parliamentary Debates*, Legislative Assembly, 26 May 1989, 2229 (Ronald Walsh, Minister for Water Resources).
29 *Water Act 1989* s 27(1), (2).
30 Ibid s 29.
31 Ibid s 32A(1).
32 Ibid s 32.

management plan ('JMP') made by a Traditional Owner Land Management Board under a *TOSA* Traditional Owner Land Management Agreement. It does not, however, require that the JMP be implemented.

There is currently no requirement in the *Water Act 1989* for WSPA consultative committees to include Indigenous representation. Accordingly, there were no Traditional Owners on the consultative committees which prepared any of the existing WSPA management plans. Nor was there anything contained in the plans to suggest that there was any consultation with or comments made by Traditional Owners in the preparation of the drafts. Only one WSPA management plan makes any reference to Traditional Owner interests, which is minimal and does not relate to management.[33]

The almost complete absence of recognition of Traditional Owner interests or Traditional Owner representation is surprising given that the plans were approved against a largely similar background to the SWSs. The reasons for the differing approaches are not clear, but it highlights the importance of legislative recognition to ensure that Traditional Owner interests are not overlooked.

In relation to the consultative committees which relate to LTWRAs and WMSs, there is also currently no requirement for them to include Indigenous representation.

One point to be made about these three consultative committees is that there is a requirement that at least 'half of the membership must consist of persons who are owners or occupiers of land in the area'.[34] This is further defined in relation to WSPA committees to mean 'farmers who own or occupy farming land in the area'.[35] This highlights a bias towards those with existing water allocations, namely farmers, and perpetuates the structural dispossession of Traditional Owners referred to earlier.

The proposed amendments, however, will require (as far as possible) that there be at least one Indigenous representative on consultative committees for WSPAs and reviews of LTWRAs,[36] and for specified Aboriginal parties to be notified of any preparation of WSPAs and LTWRAs.[37] However, WMS consultative committees will still have no requirement for any Indigenous representation. This does not prevent an Indigenous person from being appointed, but, as experience has shown, makes it less likely.

As noted earlier, consultative committees assist in the preparation of documents that guide the management of water. They are not decision-making entities. Accordingly, we need to consider what entities have responsibility for waterway management, and to what extent there is scope for Traditional Owner participation.

33 Melbourne Water, *Diamond Creek – Water Supply Protection Area – Stream Flow Management Plan 2003* (May 2003) 25.
34 *Water Act 1989* ss 29(2)(b), 22Q(b), 214(2)(a).
35 Ibid s 29(3).
36 WCL Amendment Bill cls 17, 19.
37 Ibid cls 13, 20.

Waterway management

Waterway management is governed by part 10 of the *Water Act 1989*. Under this part, entities which have responsibility for waterway management are Authorities that have a waterway management district.[38] Authorities are defined as 'a water corporation or a Catchment Management Authority'.[39]

In relation to the scope of Traditional Owner participation, there are two aspects of water management by Authorities that need to be considered here:

1 consultation processes provided for in part 10 (or in any subordinate legislative instrument) that an Authority is required to undertake; and
2 representation of Traditional Owner interests on the Authority itself – that is, the decision-making body.

The first involves the provision of Traditional Owner views to the decision-making body (the relevant Authority) which is a function of the *Water Act 1989*; the second, as it involves participation in the actual decision-making by the relevant Authority, requires an analysis of the membership requirements of the boards of CMAs under the *Catchment and Land Protection Act 1994* (Vic) ('*CALP Act*') and of water corporations under the *Water Act 1989*.

Consultation processes

Part 10 does not specify any consultation processes that authorities must undertake; however, the environment Minister can make rules relating to procedural and consultative requirements when preparing regional waterway strategies ('RWSs') and seasonal watering proposals ('SWPs').[40] To date, no such rules have been made by the Minister. Thus, it is left up to each authority to decide what procedural and consultative processes it will use.

The analysis below will focus on the preparation of RWSs because an SWP relates only to the application and use of the Victorian Environmental Water Holdings for the area in any particular season[41] and in any event must be consistent with an RWS.[42]

An RWS is an important document dealing with the management of waterways. An authority that has a waterway management district must prepare an RWS for the purposes of performing its functions.[43] Under the proposed amendments, in performing its functions, an authority will be required to (among other things) consider opportunities to provide for Aboriginal cultural values and uses of waterways.[44]

38 *Water Act 1989* s 185(1).
39 Ibid s 3 (definition of 'Authority').
40 Ibid s 197A(1)(a), (d).
41 Ibid s 192A(1).
42 Ibid s 192A(2)(a).
43 Ibid s 190.
44 WCL Amendment Bill cl 44.

In preparing an RWS, an authority must take into account any relevant strategy or statement of policy or plan made under various listed Acts.[45] Under the proposed amendments, an authority will also be required to take into account agreements made under the *TOSA* and the *AHA*, as well as any native title determination relating to the area of the RWS.[46]

All CMAs have produced a final RWS. All the RWSs refer in some form to Traditional Owner interests in waterways,[47] with most of them setting out the consultations that were undertaken with Traditional Owners. Accordingly, in the absence of any legislative requirement, it would appear that most, if not all, of the CMAs catered for some form of consultation with Traditional Owners in developing their RWSs (albeit to varying extents), which has resulted in their interests being recognised in each of the RWSs (albeit to varying degrees).

Mention should be made of regional catchment strategies ('RCSs'), also prepared by CMAs, but pursuant to the *CALP Act* [48] rather than the *Water Act 1989*. These strategies sit above RWSs in the policy hierarchy, the RWS being described by one CMA as a key sub-strategy of its catchment strategy.[49] They are more holistic in their coverage than RWSs, focussing on integrated management of land and water resources, and thus align more closely with Indigenous conceptions of country. All CMAs have an RCS in place for 2013–19. There is no requirement in the *CALP Act* to consult with Traditional Owners in the preparation of an RCS,[50] but in that regard Traditional Owners are noted as having participated in the development of most, but not all, of the existing RCSs.[51] Nonetheless, Traditional Owner interests, cultural heritage sites and values are noted in all, and all envisage that Traditional Owners will be an implementation partner, or otherwise involved in delivering the objectives of the RCS (although, as with the RWSs, the nature and extent of involvement is variable across the RCSs).

45 Ibid s 190(3)(b).

46 WCL Amendment Bill cl 45.

47 See, eg: Corangamite CMA, *Corangamite Waterway Strategy 2014–2022* (2018), ss 1.3.5, 3.2.2; East Gippsland CMA, *East Gippsland Waterway Strategy 2014–2022* (2013) s 2.2.1–2.2.2; Glenelg Hopkins CMA, *Glenelg Hopkins Waterway Strategy 2014–2022* (2014), s 1.4; Mallee CMA, *Mallee Waterway Strategy 2014–22* (2014), 17, 31; North East CMA, *North East Waterway Strategy 2014* (September 2014), 38; Melbourne Water, *Healthy Waterways Strategy – A Melbourne Water Strategy for Managing Rivers, Estuaries and Wetlands* (November 2013) 64; Wimmera CMA, *Wimmera Waterway Strategy 2014–2022* (2014), ch 6.8; Goulburn Broken CMA, *Goulburn Broken Regional Waterway Strategy 2013–2021* (2014) ch 2.7; North Central CMA, *2014–22: North Central Waterway Strategy* (2014) 13–14.

48 *CALP Act* pt 4 div 1.

49 *Corangamite Waterway Strategy 2014–2022* (2018) 42.

50 See *CALP Act* sch 2 cl 2.

51 The North East and North Central RCSs are silent on consultation with Indigenous groups. The Port Phillip and Westernport RCS is ambiguous on consultation with Indigenous groups, noting only that Traditional Owners had provided statements as a welcome to country and introduction to the strategy.

The preceding analysis indicates that CMAs did make an effort to consult with Traditional Owners in the preparation of the various strategies that they are required to prepare, even though there was no legislative requirement to do so. But the fact that the consultation was variable, resulting in varying levels of recognition in the ensuing strategies, suggests that a legislative requirement to consult may facilitate better outcomes for Traditional Owners, particularly for those whose participation in the development and resultant recognition was at the lower end of the scale.

The proposed amendments will require that the content of an RCS must 'have regard to Aboriginal cultural values and traditional ecological knowledge of management of land and water resources of the catchments in the region',[52] and that in the preparation of an RCS, any relevant specified Aboriginal party must be consulted (which can be by way or notification and an opportunity to make a submission).[53] Further, in preparing an RCS, a CMA will be required to take into account any agreement made under the *TOSA* and the *AHA*, and any native title determination.[54]

However, it is also important to note that participation by Indigenous people by various means in the preparation of these strategies will not necessarily translate into actual participation in water management, as they only set the strategic direction for the authority to follow. What may facilitate better participation outcomes for Traditional Owners is representation on the board of an authority involved in water management.

Representation on decision-making bodies

Catchment Management Authorities

Several Victorian Traditional Owner groups have expressed a desire to be represented on CMA boards in order to enhance their participation in the management of water resources,[55] so let us look at the role and functions of a CMA board to see if this can be achieved by that representation.

The functions of a CMA board are set out in s 16(1) of the *CALP Act*. What is important to note about the board's functions is that they operate largely at a strategic level. Although board members 'may exercise the powers of the Authority',[56] they do not generally undertake the day-to-day decision-making – that role is undertaken by the CEO and staff of the CMA. Appointees to the board are expected to contribute their experience and knowledge to the strategic direction

52 WCL Amendment Bill cl 83.

53 Ibid cl 86.

54 Ibid.

55 David Willington, *Technical Report No 1 – Indigenous Engagement Summary* (4 May 2011) 7, 9; Effective Management Solutions P/L, *Consultations with Indigenous Groups for the Development of the Draft Western Region Sustainable Water Strategy* (28 July 2009) 41.

56 *CALP Act* s 16(1)(b).

of the CMA's area of responsibility. Thus, a Traditional Owner appointed to a CMA board would not be there to speak on behalf of a particular Traditional Owner group or in relation to their specific area of country, but would be there to ensure that Indigenous interests in catchment management generally are factored in to the CMA's strategic direction, and to provide a link between the CMA board, its staff and the various Traditional Owner groups whose country falls within the CMA area. This is an important consideration for Traditional Owners wanting to be appointed to a CMA board. If this is not the role they are seeking, then having Traditional Owners employed by a CMA may be preferable. This too is an aspiration of many Traditional Owner groups,[57] and such opportunities, where they exist, are highly valued.[58]

In addition, CMA boundaries do not necessarily align with the boundaries of Traditional Owner country; there may be more than one Traditional Owner group whose country lies within the CMA area, or a Traditional Owner group's country could extend into other CMA areas.[59]

This could be an important limitation on the effectiveness of CMA board membership for Traditional Owners, particularly if their main aspiration is to participate in decisions relating specifically to their traditional country.

However, the CMA board is still responsible for ensuring that the CMA carries out its functions. Thus despite the limitations referred to above, it is still an important body on which Traditional Owners might want to have representation. So what are the functions of a CMA?

Various functions of CMAs are contained in the *Water Act 1989*, such as the preparation of RWSs discussed above; however, the main functions of CMAs (including the preparation of RCSs) are set out in the *CALP Act.* [60]

Beyond the preparation of an RCS, a major function of a CMA is to be a coordinating body for more localised management agencies and groups. Its involvement in decisions affecting water management at the local level is therefore mainly in setting priorities for funding local projects and partnering with other agencies and bodies for their delivery.

By being on the CMA board and ensuring that the CMA carries out its functions, Traditional Owners can be involved in overseeing the integrated management of country, which means a less fragmented role in water management.

Membership of CMA boards is governed by the *CALP Act.* [61] A CMA board can have up to nine members (including the chairperson),[62] who must have

57 Megan Goulding, *Northern Region Sustainable Water Strategy: Indigenous Engagement Project* (September 2009) 19, 20.

58 Ibid 19.

59 For example, the West Gippsland CMA area encompasses the traditional country of the Gunai/Kurnai people and the Wurundjeri people, both of whose country extends beyond the boundaries of the West Gippsland CMA area.

60 *CALP Act* s 12.

61 Ibid pt 2 div 4.

62 Ibid s 15(1).

experience and knowledge of various matters.[63] These, however, do not include Indigenous matters.

More than half of the members 'must be persons whose principal occupation is primary production'.[64] This gives primary producers a majority influence on the CMA board. Although the board is bound by the objectives of the *CALP Act*, and must comply with the Statement of Obligations made under both the *CALP Act* [65] and the *Water Act 1989*, [66] it could nonetheless be envisaged that decisions made by a CMA board might, however inadvertently, tend to lean towards the interests of primary producers, perhaps to the detriment of other interests, including Traditional Owner interests.

The lack of any requirement for Traditional Owners to be represented on the board is also significant. There is clearly nothing to prevent Traditional Owners from being appointed to a CMA board if they have knowledge or experience in one or more of the matters listed; however, the lack of any explicit requirement for a person with expertise in Indigenous matters relevant to the functions of a CMA gives the impression that such matters are not important or even relevant to the functions and duties of CMAs. Given the recognition of the importance of Traditional Owner participation in water management in the various strategies noted earlier, and the proposed amendments which would give Traditional Owners a voice on consultative committees, it is surprising that those proposed amendments do not extend to eligibility for CMA board membership, particularly since the proposed amendments do extend to eligibility for Traditional Owner membership on the Victorian Catchment Management Council ('VCMC') (discussed below). In addition, in the Commonwealth's *Water Act 2007*, eligibility for membership of the MDBA (discussed in Chapter 5) includes having a high level of expertise in 'Indigenous matters relevant to Basin water resources', so the concept is clearly not without precedent.

Victorian Catchment Management Council

The VCMC is another entity established by the *CALP Act*; however, it is not an authority for the purposes of waterway management. The functions of the VCMC are essentially to advise the Minister and to educate the public, but as a high-level advisory body it is likely to wield significant influence on policy development and can play an important role in educating the public about the value of Indigenous knowledge of water management.

Representation on the VCMC is limited to a maximum of ten.[67] Should the proposed amendments become law, unlike membership of CMA boards, there will be a specific requirement for the Minister to recommend for appointment

63 Ibid s 15(2).
64 Ibid s 15(3).
65 Ibid s 19E.
66 Ibid s 186A.
67 Ibid s 7(1).

'an Aboriginal person with experience and knowledge of Aboriginal cultural values and traditional ecological knowledge of management of land and water resources'.[68] There have already been at least two Traditional Owners on the VCMC.[69] This clearly had an impact; the VCMC acknowledged that the first appointment 'was instrumental in raising the awareness level of other Council members on Indigenous matters regarding land and water management'.[70] The proposed amendments will ensure that the previous successes of the VCMC are more easily replicated.

Representation on the VCMC does not enable Traditional Owners to participate directly in water management decisions, but it can help to promote Indigenous cultural values and knowledge generally in policy development and public education about water management.

Water corporations

Traditional Owners have also sought representation on water corporation boards.[71] Appointment to a water corporation board is governed by the *Water Act 1989*. Unlike the appointment of CMA board members under the *CALP Act*, the *Water Act 1989* is not specific about what knowledge and expertise is required for appointment to a water corporation board; it requires merely that the Minister 'ensure that each person so appointed has qualifications and experience that are relevant to the operations of the water corporation'.[72] The board's role, however, is like that of a CMA board – providing strategic direction to the water corporation. Accordingly, the same limitations noted above in relation to CMA board membership apply equally here to membership of a water corporation board.

The functions of most water corporations are generally focussed on water supply, sewerage, drainage and wastewater management rather than water management in general.[73] However, a number of rural water corporations, along with Melbourne Water, do have an important role in water management, having been delegated the role (among other things) of issuing licences to take and use water.[74] They are also involved in the preparation of WSPA management plans,[75]

68 WCL Amendment Bill cl 80.

69 Victorian Catchment Management Council, *A VCMC Position/Perspective Paper on Indigenous Engagement in Victoria* (2006) 6.

70 Ibid.

71 Willington, above n 55, 7; Effective Management Solutions, above n 55, 41.

72 *Water Act 1989* s 97. Pursuant to s 122E the Governor can make regulations regarding '(a) the qualifications required to be held by any person who holds a specific office or position in a water corporation'. No such regulations have been made.

73 Ibid s 92.

74 Under s 55 of the *Water Act 1989*, the power to determine applications for a licence to take and use is vested in the Minister; however, pursuant to the power of delegation in s 306, the Minister can delegate this power to water corporations.

75 Eight of the nine WSPA management plans show that water corporations had representation on the consultative committee appointed to prepare them.

and in their administration and enforcement.[76] Additionally, they develop local management plans for the management of unregulated waterways.[77] In that respect they are important players in the water management landscape.

Should the proposed amendments become law, a water corporation, in performing its functions, will be required to consider opportunities to provide for 'Aboriginal cultural values and uses of waterways'.[78]

Water corporations dealing solely with supply, sewerage and drainage may not be the preferred entities on which Traditional Owners will want to seek representation. CMAs, with their more holistic approach to water management as part of a larger integrated catchment management approach, align more closely with Traditional Owner aspirations for water management, which also follow a holistic approach, and thus may be viewed as more appropriate for Traditional Owner representation (subject to the limitations identified earlier). Nonetheless, it is important to be aware of the existence and nature of water corporations in the water management regulatory landscape as they still play an important role, and thus are still bodies on which Traditional Owners may aspire to have representation.

Victorian Environmental Water Holder

Finally, mention should be made of the Victorian Environmental Water Holder ('VEWH'), a further management entity on which Traditional Owners have sought legislatively mandated representation.[79] Established in 2010, the VEWH manages environmental water holdings for the purposes of:

> (a) maintaining the environmental water reserve in accordance with the environmental water reserve objective; and
> (b) improving the environmental values and health of water ecosystems, including their biodiversity, ecological functioning and water quality, and other uses that depend on environmental condition.[80]

To be appointed to the VEWH, a person must have knowledge of or experience in the fields of environmental management, sustainable water management, economics or public administration.[81] As with CMA boards, these eligibility requirements do not preclude the appointment of Traditional Owners, but neither do

76 In all of the WSPA management plans, the relevant rural water corporation has the duty of administering and enforcing the plan.
77 Local management plans 'capture and formalise existing rules in unregulated systems where there is no statutory management plan': *Victorian Waterway Management Strategy*, above n 6, 111.
78 WCL Amendment Bill cl 24.
79 Federation of Victorian Traditional Owner Corporations, Submission No 72 to the Office of Living Victoria, *Water Law Review*, 14 February 2014, para 3.3
80 *Water Act 1989* s 33DC.
81 Ibid s 33DF(2).

they promote it, despite the generally recognised importance of Indigenous knowledge to environmental management. However, in the 2016 Victorian Water Plan the government expressly committed to appointing an Indigenous Commissioner to the VEWH,[82] the first of whom was appointed in 2017.[83] The proposed amendments will also require that the VEWH consider opportunities to provide for Aboriginal cultural values and uses of waterways.[84]

In terms of the legal recognition of Traditional Owner rights to participate in water management, the above avenues were not enacted with Traditional Owner aspirations in mind. Although the *Water Act 1989* was enacted prior to *Mabo*, the *NTA* and the *NWI*, most of the relevant amendments were made in 2006, after the first consent determination of native title in Victoria in 2005.[85] The *CALP Act* was enacted after *Mabo* and the *NTA*, with additional relevant amendments also occurring in 2006.[86] This suggests that despite the existence of native title, which had been both a theoretical possibility under the *NTA* since 1993 and had become a practical reality in Victoria in 2005, the government was still unwilling to countenance the idea that Indigenous people should have a legally recognised role in land and natural resource management.

This suggests that if there is no political will to improve Traditional Owner representation, then without legislative recognition it becomes more difficult. The political environment now clearly favours such representation, but this is in stark contrast with the political environment during the Water Law Review of 2012–2014, when Traditional Owner views were essentially ignored.[87]

Concluding remarks

This chapter shows that until recently there was minimal legislative recognition of Traditional Owner rights in the *Water Act 1989*, that recognition being limited to Traditional Owner groups with a *TOSA* agreement, which involves only the right to take and use water rather than to participate in water management.

Proposed amendments will make significant inroads into rectifying the deficiencies of the *Water Act* in relation to water management. Most consultative committees will be required to include an Indigenous representative, and various

82 Department of Environment, Land, Water and Planning, *Water for Victoria – Water Plan* (2016) 172 (Action 10.8)

83 Lisa Neville, 'New Commissioner to Recognise Aboriginal Water Values' (Media Release, 16 November 2017).

84 WCL Amendment Bill cl 22.

85 *Clarke v Victoria* [2005] FCA 1795 (13 December 2005). The in-principle agreement for this consent determination was reached in October 2002: National Native Title Tribunal, *Wotjobaluk, Jaadwa, Jadawadjali, Wergaia and Jupagalk Native Title Determinations – What They Mean for the Wimmera Region* (Information Brochure, December 2006) 3.

86 These amendments included the introduction of new and improved governance arrangements for CMAs: Victoria, *Parliamentary Debates*, Legislative Assembly, 10 August 2006, 2803–4 (John Thwaites, Minister for Water).

87 See Chapter 3.

authorities established under both the *Water Act 1989* and the *CALP Act*, in performing their functions, will be required to consider opportunities to provide for Indigenous cultural values and uses of waterways.

However, representation on decision-making authorities is, and will remain, deficient, with both the *Water Act 1989* and the *CALP Act* providing no legislative basis to ensure that there is Traditional Owner representation on the boards of those authorities. There are also limitations on what Traditional Owners can achieve by having representation on water authority boards, so such representation should only be viewed as but one part of a wider solution for achieving Indigenous aspirations to participate in water management.

Finally, Indigenous participation (insofar as it is legislatively mandated) is limited to providing input into the strategic direction of water management, rather than participation in decision-making about water management.

What we can take from this is that political will and the right political climate are major factors in enhancing legislative recognition of Indigenous participation in water management. It is also clear that governments are reluctant (despite political will) to extend that participation to decision-making, preferring the Indigenous voice to be an advisory one at the strategic level. Nonetheless, this is still important, as it may then filter through to decision-making and pave the way for future participation at that level, beyond mere consultation as simply another stakeholder.

7 Victorian Indigenous rights legislation, environment and land management legislation and the management of water resources

Land rights legislation

Introduction

Under land rights legislation, or with land transferred pursuant to a native title settlement agreement,[1] there may be potential for Indigenous rights holders to be involved in managing the waters on that land. These would be similar to the rights of other land holders, such as by way of statutory entitlements.[2]

The popular history of land rights in Australia has largely focused on the Northern Territory. The enactment of the *Aboriginal Land Rights (Northern Territory) Act 1976* (Cth) ('*ALR(NT) Act*') in particular has been well documented as being a watershed for the legislative recognition of Indigenous land rights in Australia and is 'widely regarded as being the high water mark in land rights legislation'.[3] What is less well known is that Victoria had already legislated for grants of land to Aboriginal people in 1970, namely the *Aboriginal Lands Act 1970* (Vic) ('*ALA*'). A significant difference between the two Acts was that the *ALA* was for the grant of specific parcels of land whereas the *ALR(NT) Act* set up a Territory-wide land claims process. The other notable difference is that the *ALR(NT) Act* has resulted in relatively comprehensive Indigenous engagement in land (and water) management, but Victorian land rights laws have not had the same widespread effect for Traditional Owners in Victoria.

This section examines the extent to which land rights legislation in Victoria has been able to assist Indigenous Victorians to participate in decision-making about water management.

1 The Gunditjmara and Wimmera Clans native title settlements both involved the transfer of freehold title of parcels of land to the native title holders, totalling approximately 292.8 ha. Importantly, the Gunditjmara settlement transferred title to Lake Condah Reserve, part of the Budj Bim National Heritage Landscape.

2 These would include statutory rights of a riparian nature, as riparian rights have arguably been abolished in Victoria, see Chapter 2.

3 Heather McRae et al, *Indigenous Legal Issues, Commentary and Materials* (Lawbook, 4th edn, 2009) 223.

Victorian land rights legislation, 1970–2010

The *ALA* was enacted 'to provide for the Lands reserved for the Use of the Aborigines at Framlingham and Lake Tyers to be vested in a Framlingham Aboriginal Trust and a Lake Tyers Aboriginal Trust respectively'.[4] Land adjacent to the Hopkins River at Framlingham and land adjacent to the lake itself at Lake Tyers could be specifically excluded from the grant,[5] although the Governor in Council could grant perpetual licences to occupy the land to the respective trusts.[6] This was to 'allow Aboriginal people the unfettered use of these valuable frontages'.[7]

Further grants of specific parcels of land to Aboriginal people were made pursuant to legislation enacted in 1982, 1989, 1991 and 1992 respectively.[8] The total area of land handed back to Aboriginal people under this legislation, including the *ALA*, amounts to approximately 18.78 square kilometres, the vast bulk of this land (over 85 per cent) being the Lake Tyers grant of 1970. To further put this into context, the area of Victoria is 227,416 square kilometres,[9] and therefore the total percentage of land handed back to Aboriginal people in Victoria up to and including 1992 via Victorian land rights legislation was approximately 0.008 per cent of the state. So there was very little land actually transferred to Aboriginal people pursuant to these Acts over which they could assert their traditional rights to manage the water resources on that land.

Land granted under these Acts was generally not granted to the Traditional Owners, or entities representing the Traditional Owners. Title to Lake Tyers and Framlingham, for example, was granted to trusts representing the residents of those reserves,[10] which included but was not limited to the Traditional Owners of the area.[11] Given that it is the concept of Indigenous sovereignty that underpins Indigenous people's claims to manage the waters on their lands, the granting of land to entities which include Indigenous people who do not have a traditional connection with the land can be seen as a serious weakness in Victoria's land rights legislation during this time. This is exacerbated by the fact that some of those who *do* have a traditional connection (but were non-resident at the time of the grant) are excluded from membership of the land-holding entity.

4 *ALA* long title.
5 Ibid s 9(4).
6 Ibid s 9(5).
7 Victoria, *Parliamentary Debates*, Legislative Assembly, 28 October 1970, 1421 (E R Meagher, Minister for Aboriginal Affairs).
8 *Aboriginal Land (Aborigines' Advancement League) (Watt Street, Northcote) Act 1982* (Vic); *Aboriginal Land (Northcote Land) Act 1989* (Vic); *Aboriginal Land (Manatunga) Act 1992* (Vic); *Aboriginal Lands Act 1991* (Vic).
9 Geoscience Australia <www.ga.gov.au/education/geoscience-basics/dimensions/area-of-australia-states-and-territories.html>.
10 *ALA* s 3.
11 It appears that this was not actually the intention of the government at the time; see Victoria, *Parliamentary Debates*, Legislative Assembly, 28 October 1970, 1419 (E R Meagher, Minister for Aboriginal Affairs).

The land granted under the 1982 and 1989 Acts was to the Aborigines' Advancement League ('AAL'), an organisation representing all Aboriginal people in Victoria. These were small parcels of land in metropolitan Melbourne, so the land itself was not of a size or nature or in a location for which Traditional Owners would be seeking management rights over water. Like other owners of freehold property in metropolitan Melbourne, the AAL would currently have a retail water supply company supplying water purchased from Melbourne Water, one of the 16 state-owned water businesses governed by the *Water Act 1989* (Vic) ('*Water Act 1989*').[12]

The land grants made in 1991 consisted of three former cemeteries that had been part of the Coranderrk, Ebenezer and Ramahyuck missions. In the case of Ebenezer and Ramahyuck, the land was granted to the local Aboriginal Cooperative for the area, membership of which was open to Aboriginal residents of the area as well as non-Aboriginal spouses, de-facto spouses, widow(er)s or parents of Aboriginal residents.[13] The area of land encompassed by the three cemeteries is approximately 2.7, 3.9 and 8.6 hectares respectively, none of them abutting a water course.

The land granted to the Murray Valley Aboriginal Co-operative Limited under the 1992 legislation is of a similar nature to the three cemetery parcels.

Finally, the Commonwealth also passed legislation granting land to Aboriginal Victorians, namely the *Aboriginal Land (Lake Condah and Framlingham Forest) Act 1987* (Cth). This legislation was enacted by the Commonwealth, rather than Victoria, because the Victorian Labor government at the time had 'repeatedly tried to pass limited land rights and cultural heritage legislation but was blocked by the Legislative Council in which the Opposition parties had the majority'.[14] Thus, to circumvent this deadlock, the state government requested that the Commonwealth enact a limited version of its proposed legislation using its powers under s 51(xxvi) of the *Commonwealth Constitution*.

12 Department of Sustainability and Environment, *A Governance Guide to the Victorian Water Industry* (Victorian Government, East Melbourne, 2011).

13 See, eg, clause 8(2) of 'Annexure A: Draft Rules of Gippsland and East Gippsland Aboriginal Co-Operative Limited. Registered as a non-trading Co-operative with shares under the Co-operatives Act 1996 (Vict.)', copy on file with the author. Ebenezer Mission was granted to Goolum Goolum Aboriginal Co-operative, Ramahyuck was granted to the Gippsland and East Gippsland Aboriginal Co-operative. Coranderrk was granted to Wurundjeri Tribe Land and Compensation Cultural Heritage Council (WTLCCHC), the only organisation whose membership appears to be limited to Traditional Owners: *Decision of the Victorian Aboriginal Heritage Council in Relation to an Application by Wurundjeri Tribe Land and Compensation Cultural Heritage Council Inc to be a Registered Aboriginal Party* (22 August 2008) <www.vic.gov.au/aboriginalvictoria/heritage/registered-aboriginal-parties/wurundjeri-land-and-compensation-cultural-heritage-council-aboriginal-corporation.html>.

14 Garth Nettheim, Gary D Meyers and Donna Craig, *Indigenous Peoples and Governance Structures: A Comparative Analysis of Land and Resource Management Rights* (Aboriginal Studies Press, 2002) 262. See also Jessica Weir, *The Gunditjmara Land Justice Story* (Native Title Research Unit, AIATSIS, 2009) 11.

Some 1,183 hectares comprising parcels of land at Condah and Framlingham Forest were handed back to Aboriginal corporations.[15] These corporations were given significant management powers over these reserves, including the power to make by-laws.[16] However, none of the matters in the by-law-making power are specifically in relation to water resources.

Like much of the land handed back under Victorian land rights legislation, the Condah parcels consisted of the mission and cemetery. However, unlike the transfer of many other cemetery and mission titles to Aboriginal corporations, the corporation to which the Condah titles were granted did represent the Traditional Owners.[17]

One of the two Condah parcels abuts Darlot Creek, and the Framlingham Forest parcel abuts the Hopkins River. But unlike the *ALA*, the Commonwealth Act does not exclude from grant the land adjacent to these water courses.

Clearly, land rights legislation in Victoria was not intended to have any effect on the existing water management regime, and any ability to manage water over these parcels would be very limited. The only water rights that the Aboriginal cooperatives and councils could have over these parcels of land would be those of an ordinary landholder or occupier under the *Water Act 1989*, such as the right to take and use water for domestic and stock purposes, and the general participatory rights discussed in the previous chapter.

This would also be the case for other land owned and/or managed by Traditional Owners, such as the land identified in the SAMLIV Report of 2003, the *Strategy for Aboriginal Managed Lands in Victoria*, [18] and land purchase by the Indigenous Land Corporation ('ILC'),[19] generally later transferred in title to Traditional Owners.

Historically, the fragmentary nature, grant size, recipient status and scope of land rights legislation in Victoria has meant that it was not a viable avenue by which Indigenous people were able to assert their rights as Traditional Owners to participate in managing water resources on land granted pursuant to that legislation.

It is only recently that the Victorian government enacted legislation with statewide application, namely the *Traditional Owner Settlement Act 2010* (Vic) ('*TOSA*'), which created a mechanism by which Victorian Traditional Owners can seek to have title of some of their traditional lands transferred to them, and by

15 The Condah parcels were transferred to the Kerrup-Jmara Elders Aboriginal Corporation and the Framlingham Forest parcel to the Kirrae Whurrong Aboriginal Corporation.

16 *Aboriginal Land (Lake Condah and Framlingham Forest) Act 1987* (Cth) ss 15, 23.

17 Weir, above n 14, 14. The Kerrup-Jmara Elders Aboriginal Corporation is no longer in existence and the land has now been transferred to the Gunditj Mirring Aboriginal Corporation, the native title prescribed body corporate.

18 SAMLIV Project Team, *Strategy for Aboriginal Managed Lands in Victoria* (SAMLIV Steering Committee, December 2003).

19 Established in 1995, the ILC was an initiative of the Keating Labor government in response to the decision in *Mabo*.

which the management of water resources may be a possibility. As the most important development in Victoria in relation to the recognition and regulation of Indigenous rights and interests, it is therefore that Act to which this analysis now turns.

Victorian Traditional Owner Settlement Act 2010 *(Vic)*

Background

In 2005, a group of Traditional Owners formed the Victorian Traditional Owner Land Justice Group ('VTOLJG') and called for a comprehensive land justice settlement to 'finally address the fundamental need for land justice in Victoria'.[20] Land justice included ownership and management of land and natural resources that was not dependent on recognition of native title,[21] with a particular mention of water,[22] all in an environment of informed consent.[23] The VTOLJG produced a detailed discussion paper in August 2006 expanding on the matters (including water)[24] outlined in 2005, which was provided to the state in September that year. These developments formed the basis for negotiations between Victorian Traditional Owners, represented by the VTOLJG, and the state on a new policy framework for land justice in Victoria. These negotiations were resourced by the state and Native Title Services Victoria ('NTSV').

In March 2008 a Steering Committee was set up to investigate and develop options for policy reform in relation to land justice in Victoria, including in relation to native title.[25] It produced its final report ('*Steering Committee Report*')[26] in December 2008 which was subsequently endorsed by the state government in June 2009.[27]

The *Steering Committee Report* contained numerous recommendations for land justice, referred to in the report as 'Core Principles'. In relation to water, Core

20 VTOLJG, *Statement* (17 February 2005), cl 1.1 <www.landjustice.com.au/document/Communique-Statewide-Meeting-17-18Feb05.pdf>.

21 Ibid cl 3.

22 Ibid cl 9.

23 Ibid cl 2.

24 VTOLJG, *Towards a Framework Agreement Between the State of Victoria and the Victorian Traditional Owner Land Justice Group* (Discussion Paper, 26 August 2006) cl 7.2 <www.landjustice.com.au/document/Framework-Discussion-Paper-26-August-2006.pdf>.

25 The Steering Committee was chaired by Professor Mick Dodson and comprised of members of the VTOLJG, NTSV and senior departmental representatives of the state government. The author was an employee of NTSV during this time and participated in one of the working groups which were set up to advise the Steering Committee on various topics contained in the terms of reference and to assist in negotiations.

26 Steering Committee for the Development of a Native Title Settlement Framework, *Report of the Steering Committee for the Development of a Native Title Settlement Framework* (December 2008).

27 Rob Hulls, 'Keynote Address' (Speech delivered at the AIATSIS Native Title Conference, Melbourne, 4 June 2009).

Principle #31 recommended increased participation by Indigenous people in natural resource management, including a statutory requirement to consult.[28] Core Principle #32 recommended (among other things) statutory recognition of customary non-commercial use and access to water, and greater consideration of Traditional Owner groups' interests in developing and implementing water resource actions. As part of its response to the recommendations contained in the *Steering Committee Report*, the Victorian government enacted the *TOSA*.

Unfortunately for Victoria's Traditional Owners, the recommendations contained in Core Principles #31 and #32 were not incorporated into the *TOSA*. So what did get included in the *TOSA*?

The Traditional Owner Settlement Act *as it relates to Indigenous water rights*

Section 9 of the *TOSA* provides for recognition of Traditional Owner rights and interests akin to native title rights, with one glaring exception, namely water rights. Water rights are not mentioned, but are instead a subset of natural resource rights in s 9. To have Traditional Owner rights and interests recognised, the Traditional Owner group must enter into a recognition and settlement agreement ('RSA').[29] The exercise of natural resource rights is then dealt with in pt 6. When the *TOSA* was first enacted, in order to exercise any natural resource rights a traditional owner group entity ('TOGE')[30] was required to first enter into a natural resource agreement ('NRA')[31] after which it could then obtain an authorisation order.[32]

The provisions in relation to authorisation orders were repealed in November 2016[33] in recognition that they 'fall short of the natural resource rights available under the [*NTA*]'.[34]

Water resource rights are currently limited to taking and using water 'for traditional purposes'.[35] Traditional purposes are defined as the purposes of providing for any personal, domestic or non-commercial communal needs of the members of the Traditional Owner group,[36] terminology which is consistent with s 211(2)(a) of the *NTA*.[37]

28 Core Principle #31 did not specify water, but water was clearly intended to be dealt with as a natural resource, as evidenced by Core Principle #32. This is also apparent from the *TOSA*.
29 *TOSA* s 9.
30 Ibid s 3 (definition of 'traditional owner group entity').
31 Ibid s 80.
32 Ibid s 85(1).
33 *Traditional Owner Settlement Amendment Act 2016* (Vic).
34 Victoria, *Parliamentary Debates*, Legislative Assembly, 31 August 2016, 3236 (Martin Pakula, Attorney-General). For more on how authorisation orders operated, see Katie O'Bryan, 'Victoria's Traditional Owner Settlement Act and Indigenous Management of Water Resources: An Improvement on the Native Title Act or More of the Same?' (2016) 8(26) *Indigenous Law Bulletin* 21, 23.
35 *TOSA* s 84(a). Water is not one of the natural resources encompassed by s 84(b).
36 Ibid s 79.
37 It is also consistent with determinations of native title made in Victoria. Note, however, that in relation to s 211 of the *NTA*, s 211(3) includes cultural and spiritual activities as a class of activities to which the section applies.

So, like native title rights to water, water rights under the *TOSA* are limited to rights to take and use water; they do not include management rights.

Consultation and procedural rights – the Land Use Activity regime

Insofar as Indigenous water rights encompass decision-making or consultation rights in relation to the management of water resources, there may be some scope for Traditional Owners to have them included in a *TOSA* settlement package via the Land Use Activity regime ('LUA regime').

OVERVIEW

The LUA regime is intended to be generally equivalent to the Future Act regime in the *NTA* (albeit simplified).[38] For the LUA regime to apply, a TOGE must first enter into a land use activity agreement ('LUAA').

A LUAA reached under the LUA regime provides procedural rights in relation to land use activities occurring on land recognised under an RSA as being Traditional Owner land. The nature of the procedural rights of Traditional Owners depends on the nature of the activity being undertaken. Activities can be categorised as (in increasing order of impact) routine, advisory, negotiation (Class A or Class B) or agreement activities. In general, the category into which each activity is placed is negotiated by the parties to the LUAA, namely the Traditional Owners and the state. Importantly, as the following analysis will show, activities regarding water management and allocation to which the LUA regime applies are very limited, and therefore the LUA regime can be seen as largely only an indirect avenue for participating in the management of water resources.[39]

A routine activity requires no involvement of Traditional Owners before it can proceed. This type of activity includes an exploration activity where the proponent has agreed to comply with pro-forma conditions.[40] An advisory activity requires that the TOGE be notified of the proposal to carry out the activity and to be consulted about the proposal.[41] If an activity is categorised as a negotiation activity, the parties must negotiate in good faith towards reaching an agreement about the carrying out of the proposed activity.[42] If agreement cannot be reached within six months, then the matter can be taken to the Victorian Civil and Administrative Tribunal ('VCAT') for a determination.[43] The Minster can also make a determination and substitute it for the determination by VCAT if it is in the public

38 Victoria, *Parliamentary Debates*, Legislative Assembly, 28 July 2010, 2753 (John Brumby, Premier).

39 The exception to this would appear to be the preparation of draft management plans under s 31 of the *Water Act 1989*, discussed in Chapter 6.

40 *TOSA* s 33(2).

41 Ibid s 34.

42 Ibid s 50.

43 Ibid s 53.

interest.[44] Traditional Owners have no right of veto over activities classed as negotiation activities. There are also limits to what activities can be included in this category. A negotiation activity (Class A) is limited to a 'significant land use activity', and a negotiation activity (Class B) is limited to an activity which is either a 'significant land use activity' or 'limited land use activity'. The final category, agreement activities, can only include an activity which is a 'significant land use activity'. However, the consent of the TOGE is required for an agreement activity to proceed, and there is no review by VCAT if agreement cannot be reached. The agreement category thus provides the strongest procedural rights of all the categories.

APPLICATION TO WATER MANAGEMENT

It is now necessary to set out in more detail a few of the defined terms noted above. This will assist in understanding how Traditional Owners might be able to use the LUA regime to ensure that they are consulted about proposed activities that relate to or may affect water resources in the area of their agreement. These definitions place upper limits on which category each activity can be allocated to; it may be that an activity will be allocated to a lower category, depending on the outcome of negotiations.

The definition of 'land use activity' relevantly includes:

- the granting of a 'public land authorisation' over land, including amendments/variations which allow for the change of the authorised activity;[45]
- the preparation of a management plan under ss 17, 17B, 17D or 18 of the *National Parks Act 1975* (Vic);[46] and
- the preparation of a draft management plan under s 31 of the *Water Act.* [47]

The definition of 'public land authorisation' referred to above includes 'a licence under Division 2 of Part 5 of the *Water Act 1989* to construct any works on a waterway or bore'.[48] A 'limited land use activity' is defined to include 'a land use activity that is for the purpose of the establishment, use or operation of any specified public works'. Specified public works are also defined and include:

- a well or bore for obtaining water;
- a pipeline or other water supply or reticulation facility;

44 Ibid ss 65, 66.
45 Ibid s 28(a)(i), (ii).
46 Ibid s 28(l). A management plan made pursuant to the *National Parks Act 1975* (Vic) may be over an area which contains water resources. If it is not already the subject of a joint management plan made by a TOLMB via a TOLMA, then this is potentially an alternative avenue for input, albeit only in the advisory category.
47 Ibid s 28(p). These are water supply protection area management plans, discussed in Chapter 6.
48 Ibid s 27(1) *public land authorisation* ... (h).

- a drainage facility, or a levee or other device for the management of water flows; and
- an irrigation channel or other irrigation facility.[49]

Finally, a 'significant land use activity' is defined as including:

- an authorisation to occupy, use or otherwise access the land for ten years or more;
- Carbon Land Sequestration Agreements;
- the clearing of the land or the carrying out of any works on the land which has a substantial impact on the physical quality of the land, having regard to the size and scale of the activity;
- various types of earth resource and infrastructure authorisations; and
- grants of land in freehold under the *Land Act 1958* (Vic).[50]

Accordingly, the highest category that a TOGE could negotiate for any of the specified public works noted earlier to be categorised is negotiation activity (Class B). Then any proponent who wished to undertake one of those public works – for example, construct a device for the management of water flows – would be required to negotiate in good faith with the TOGE in order to reach agreement about *how* the activity can proceed, although not on *whether* it can proceed.

A TOGE could also negotiate for the relevant public land authorisation (ie, construction of any works on a waterway or bore) or other relevant land use activity (ie, preparation of a draft management plan under s 31 of the *Water Act*) to be categorised as an advisory activity, in which case it would be required to be notified of the proposed activity and be entitled to be consulted about the effects of the proposed activity on Traditional Owner rights. This is the highest category to which these activities can be allocated.

Given the definition of significant land use activity, it appears that no land use activity that is directly related to water can be categorised as an agreement activity. However, negotiations in relation to proposed activities that have been categorised as agreement activities may be able to consider impacts on water resources provided that those impacts relate to Traditional Owner rights.[51] As water rights are not specifically included in the list of Traditional Owner rights able to be recognised pursuant to s 9, they could arguably fall within s 9(f) or (h), which relate respectively to 'the ability to take natural resources on or depending on the land' and 'the protection of places and areas of importance on the land'. An example of s 9(f) might be where an agreement activity (such as 'the clearing of the land or the carrying out of any works on the land which has a substantial impact on the physical quality of the land')[52] also impacts on a water resource authorisation, or,

49 Ibid s 27(1) (definition of 'specified public works').
50 Ibid s 27(1) (definition of 'significant land use activity').
51 Ibid s 50(2)(a).
52 Ibid s 27(b) (definition [in part] of 'significant land use activity').

for s 9(h), on a water body which is also a place or area of importance to the Traditional Owner group.

There is no equivalent in the LUA regime of s 24HA of the *NTA*, which contains the limited right to be notified and be given the opportunity to comment on activities relating to the management and regulation of water. This has both benefits and disadvantages. One benefit is that it enables various water-related activities to potentially be categorised as negotiation activities, which is not possible under s 24HA of the *NTA*. The disadvantage is that any water-related activities that do not fit within the definition of 'land use activity' or other relevant definition are not able to be included in any of the categories, and therefore Traditional Owners will have no procedural rights at all in relation to those activities; they are effectively deemed routine activities. So decisions regarding the allocation of water licences, for example, would fall into this category.

In contrast, although s 24HA of the *NTA* is a very limited right, somewhat analogous to the 'advisory' category in the LUA regime, it at least requires that native title claimants/holders be notified of and given an opportunity to comment on all activities related to the management and regulation of water (with the exception of the making, amendment or repeal of water management legislation). Section 24HA also requires the payment of compensation for any effects on native title rights and interests caused by the activity, in contrast with the *TOSA*, whereby routine and advisory activities do not attract the ability for Traditional Owners to negotiate any 'community benefits' for impacts on their Traditional Owner rights. Negotiation and agreement activities may, however, provide for community benefits.[53]

A further issue to consider here is whether a failure to comply with the procedural requirements to notify and consult with the relevant TOGE will render the authorisation of the proposed activity invalid. In that regard, s 35 of the *TOSA* states that

> [a] decision maker who is carrying out an advisory activity on any agreement land to which a direction under section 34 applies [being a direction to the decision-maker to notify and consult the TOGE], must give effect to the direction in carrying out the activity.

In relation to s 24HA of the *NTA*, a majority of the Federal Court (in obiter) was of the view that a failure to comply with procedural requirements in that section would not render an authorisation invalid.[54] Reasons given by the court in support of this view included the specific reference to invalidity for non-compliance in relation to other Future Act provisions of the *NTA*, the express statement in the Explanatory Memorandum that failure to comply would not result in invalidity, and by looking at the language of the provision, and the scope and objects of the legislation as whole (the *Project Blue Sky* test).[55]

53 See, eg, ibid ss 40(1), (4), 40A.
54 *Lardil Peoples v Queensland* (2001) 108 FCR 453, [117].
55 *Project Blue Sky v Australian Broadcasting Authority* (1998) 194 CLR 355.

However, should this question ever be brought before the court, factors that point to a different result (ie invalidity) in relation to a failure to comply with s 35 of the *TOSA* include:

- section 36, which states that '[a] direction of the Minister has effect despite anything to the contrary in the Act or regulations under which the land use activity is being carried out';
- the arguably stronger right of consultation in contrast to the opportunity to comment;
- the inability for Traditional Owners to obtain community benefits (ie compensation) for any impact of advisory activities on their Traditional Owner rights;
- the lack of reference in the Explanatory Memorandum that a failure to comply is not intended to result in invalidity; and
- the scope of the application of the procedural requirements for advisory activities – that is, they are not limited to activities in relation to the management or regulation of water.

On the other hand, the categorisation of an activity as an advisory activity is the outcome of negotiations, not of statute, and therefore presumably the Traditional Owners would be aware of the potential limitations of the procedural requirements for advisory activities.

This may all be moot because, as just noted, Traditional Owner groups still need to negotiate to have any land use activities affecting water resources included as a land use activity in any LUAA reached pursuant to the *TOSA*. Given that only specified land use activities can be included in a LUAA, with the only specified land use activity directly relating to the management of water being the preparation of draft management plans pursuant to s 31 of the *Water Act*, it appears that apart from this one activity, the LUA regime can only be used indirectly by Traditional Owners to participate in water management. The LUA regime, therefore, has little to offer Traditional Owners in terms of meaningful participation in water management.

An alternative could be to look to other means of being involved in water management, such as via joint management.

Joint management

Joint management in the Victorian context has been described as 'a formal partnership arrangement between Traditional Owners and the State where both share their knowledge to manage specific national parks and other protected areas'.[56]

56 Parks Victoria, Aboriginal Joint Management <http://parkweb.vic.gov.au/park-management/aboriginal-joint-management>. Parks Victoria is a statutory authority established pursuant to the *Parks Victoria Act 1998* (Vic).

Background

Joint management in Victoria commenced with the enactment of the *Parks and Crown Land Legislation Amendment (River Red Gums) Act 2009* (Vic) ('*River Red Gums Act*'). The *River Red Gums Act* amended various Acts, but importantly it amended the *Conservation Forests and Lands Act 1987* (Vic) ('*CFL Act*'). Under agreements reached pursuant to these legislative amendments, known as traditional owner land management agreements ('TOLMAs'), a joint management board with a majority of Traditional Owners (a Traditional Owner Land Management Board ('TOLMB')) can be established to manage an agreed area of significance to Traditional Owners.

The *River Red Gums Act* was enacted to fulfil a commitment by the Labor state government 'to involving traditional owners in the management of public land and to helping traditional owners achieve their long-held aspirations to be involved in caring for country'.[57] It was initially intended to result in agreements for management arrangements over the Barmah National Park with the Yorta Yorta People and the Nyah-Vinifera Park with the Wadi Wadi People, but was designed so that agreement could be reached over other areas of land and with other Traditional Owner groups in the future.[58]

It was the government's view at the time that the *River Red Gums Act* would enhance Traditional Owner involvement in decision-making over their traditional lands, the Minister stating: 'This is historic legislation for Victoria. For the first time, the ability for traditional owners to be decision-makers and to have a substantial involvement in the management of their traditional lands will be enshrined in the state's law.'[59]

Joint management and the *Traditional Owner Settlement Act*

The *TOSA* then took this a step further in two respects. First, by amending the *CFL Act* to provide for the making of joint management plans ('JMPs') by TOLMBs.[60] Associated with this were amendments to various other Acts to require that any person responsible for the management of appointed land under those Acts must ensure that management is not inconsistent with any JMP for that land.[61]

Contrast this with the cooperative management agreements reached prior to the enactment of the *TOSA* (and the *River Red Gums Act*). Under these cooperative management agreements, advisory councils were set up comprised of Traditional

57 Victoria, *Parliamentary Debates*, Legislative Assembly, 15 October 2009, 3637 (Peter Batchelor).

58 Ibid. At the time of writing, the Yorta Yorta People had reached agreement, along with the Gunaikurnai and Dja Dja Wurrung Peoples as part of their respective *TOSA* settlements.

59 Ibid.

60 *CFL Act* pt 8A div 5A.

61 *Crown Land Reserves Act 1978* (Vic) s 20A; *Forests Act 1958* (Vic) s 57A; *Land Act 1958* (Vic) s 4C; *National Parks Act 1975* (Vic) s 16B; *Wildlife Act 1975* (Vic) s 18B.

Owners and state representatives to provide non-binding advice to park managers, but which had no power to prepare a JMP. The first such agreement was with the Yorta Yorta People. Notably as far as water bodies are concerned, it includes 'Kow Swamp, Barmah State Park, Barmah State Forest, and public land and waters along sections of the Murray and Goulburn Rivers'.[62] A cooperative management agreement was also reached with the Wimmera Clans over various Crown land reserves within the external boundary of their consent determination, which includes the Wimmera River, Lake Hindmarsh and Lake Albacutya,[63] and with the Gunditjmara over Mt Eccles National Park.[64]

The second important development of the *TOSA* was to allow for agreements to be made which provide for the transfer of title of areas of reserved public land to TOGEs on the condition that the land will be jointly managed on the same basis for which it had been reserved prior to the transfer.[65] For example, a national park would continue to be managed as a national park and a nature reserve would continue to be managed as a nature reserve. This title is known as Aboriginal title. A condition for an agreement to transfer Aboriginal title is that the TOGE enter into a TOLMA with the relevant Ministers under s 82P of the *CFL Act*. At the time of writing, there had been only two such instances of concurrently reached agreements under the *TOSA*, namely the agreements reached in settlement of the Gunaikurnai native title claim in 2010 and in settlement of the Dja Dja Wurrung native title claim in 2013.

The total amount of land transferred to Traditional Owners in Aboriginal title to date via the *TOSA* is approximately 93,000 ha.[66] Given that this land relates to the traditional country of only two of the many Traditional Owner groups in Victoria, with many groups still to take advantage of the *TOSA*, it is a significant (albeit still inadequate) improvement on the amount of land transferred prior to 1992 under the various land rights statutes.

But does this translate into an ability of Traditional Owners (via their nominated TOGE) to participate in the management of the water bodies on or flowing through the land that is the subject of a land agreement reached under the *TOSA*?

Aboriginal title under the *TOSA* and associated joint management (ie TOLMA) arrangements do not automatically allow for management of the water bodies on

62 Department of Justice and Department of Sustainability and Environment, *Yorta Yorta Co-operative Management Agreement* (Fact Sheet, May 2004) 1.

63 *Co-operative Management Agreement between Barengi Gadjin Land Council and State of Victoria* (13 December 2005) Schedule 2.

64 National Native Title Tribunal, *The Gundtijmara People's Native Title Determination* (30 March 2007).

65 *TOSA* s 12(3), s 20.

66 Approximately 47,500 ha were transferred to the Dja Dja Wurrung: *Settlement Between the Dja Dja Wurrung Traditional Owner Group and the State of Victoria* (Fact Sheet, 28 March 2013); and approximately 45,500 ha were transferred to the Gunaikurnai: Gavin Jennings on behalf of the State of Victoria, Gunaikurnai Land and Waters Aboriginal Corporation and Secretary to the Department of Sustainability and Environment, *Traditional Owner Land Management Agreement* (22 October 2010) ('Gunaikurnai TOLMA'), copy on file with the author.

the subject land. These may have to be specifically negotiated for inclusion as part of the agreement.[67] An example of where negotiations failed to achieve the inclusion of a water body is the Lake Tyers Catchment Area, transferred in Aboriginal title to the Gunaikurnai People as part of their native title settlement.[68] Joint management arrangements, even when they do include water bodies, may not encompass them in their entirety. This can lead to a fragmentation of management responsibilities. Another example from the Gunaikurnai settlement is the Mitchell River, which flows through the Mitchell River National Park. It is only the section of the Mitchell River that flows through the National Park that is subject to the Gunaikurnai TOLMA and for which the TOLMB can therefore prepare a JMP.

The content of a JMP is set out in the relevant TOLMA.[69] The Gunaikurnai TOLMA, for example, provides that '[a] Joint Management Plan must provide for the sustainable management of the Appointed Land and may include strategies for: …'[70] Although there is no specific reference to water management as one of the topics for which strategies may be developed, the term 'may include' is inclusive rather than exhaustive. This suggests that other strategies could also be included in the JMP, which further suggests that water management strategies are not precluded. Indeed, given that the Gippsland Lakes Coastal Park[71] is one of the areas transferred in Aboriginal title to the Gunaikurnai and for which the Gunaikurnai TOLMB has been given responsibility to prepare a JMP, it would be bizarre if a JMP encompassing this park could not include strategies for water management.

Proposed amendments to Victoria's *Water Act* will require Sustainable Water Strategies and regional waterway strategies to take into account RSAs and any further agreement made under the *TOSA*, [72] which would include TOLMAs and therefore JMPs made pursuant to TOLMAs, but this does not require that a JMPs made pursuant to a TOLMA be implemented.

The *Water Act* in that respect can be contrasted with s 20A of the *Crown Land (Reserves) Act 1978* (Vic) which provides:

> If any appointed land of a Traditional Owner Land Management Board constitutes the whole or a part of land reserved under this Act, the person responsible for the management of that appointed land under this Act must ensure that the land is managed in a way that is not inconsistent with any joint management plan for the land.

67 Katie O'Bryan, 'The National Water Initiative and Victoria's Legislative Implementation of Indigenous Water Rights' (2012) 7(29) *Indigenous Law Bulletin* 24, 26.
68 Lake Tyers itself was not included in the transfer of Aboriginal title, merely the land in the catchment area surrounding the Lake. This was despite the recognition of native title over both the Lake and the surrounding catchment area.
69 *CFL Act* s 82PC.
70 Gunaikurnai TOLMA, above n 66, cl 3.3.
71 The Gippsland Lakes Coastal Park encompasses Lake Reeve.
72 Water and Catchment Legislation Amendment Bill 2017 (Vic) cls 7, 45.

Similar provisions are found in the *Forests Act 1958* (Vic),[73] the *Land Act 1958* (Vic),[74] the *National Parks Act 1975* (Vic)[75] and the *Wildlife Act 1975* (Vic),[76] all of which were inserted as the result of amendments made by the *TOSA*.

Despite assertions of the significant benefits to Traditional Owners of the *TOSA*, [77] it is still inadequate insofar as it facilitates Indigenous involvement in the management of water resources.

Concluding remarks

Traditional Owners have obtained ownership of land via a variety of different avenues in order to manage the land and the resources on that land, including water resources. But land title/ownership, without more, is generally insufficient to enable participation in managing water resources, other than what is currently available under the *Water Act*.

Traditional Owners with recognised rights over land under the *TOSA* can reach a LUAA with the state about certain procedural rights which may relate to the management of water resources on that land. However, because these are negotiated rights and are limited to specified land use activities (virtually none of which directly relate to water management), the LUA regime does not have much to offer Traditional Owners.

The *CFL Act* (as amended by the *River Red Gums Act* and *TOSA*) provides the most potential for Traditional Owner participation in water management, without even the need for Traditional Owners to have title. But TOLMAs reached under the *CFL Act* apply only to public land, or Aboriginal title land (and therefore only water bodies on that land), so they too are limited in their application and have the capacity to fragment Indigenous management of water.

Heritage protection legislation: the *Aboriginal Heritage Act 2006* (Vic)

Background

Aboriginal Victorians rely on two Acts to protect their cultural heritage. Initially, the first was the *Archaeological and Aboriginal Relics Preservation Act 1972* (Vic) ('*AARP Act*') which was intended to protect, as the name suggests, Aboriginal relics along with archaeological areas, being essentially those areas on which relics were found. The second Act is Commonwealth legislation, namely the *Aboriginal and Torres Strait Islander Heritage Protection Act 1984* (Cth) ('*ATSIHP Act*') the

73 Section 57A.
74 Section 4C.
75 Section 16B.
76 Section 18B.
77 John Brumby, 'New Framework a Just Approach to Native Title' (Media Release, 28 July 2010); Victorian Traditional Owner Land Justice Group, 'Land Justice Within Reach' (Media Release, 28 July 2010).

purposes of which are to preserve and protect areas and objects of significance according to Aboriginal tradition.[78] The *ATSIHP Act* applies Australia-wide and therefore applied (and still applies) to Victoria from the time of its enactment. However, amendments were made in 1987 that were specific to Victoria in order to enhance the protection of Aboriginal cultural heritage in that state beyond the limited protection available under the *AARP Act.*

These amendments to the *ATSIHP Act* came about because the Victorian Labor government had been unable to enact its own heritage protection (and land rights) legislation, due to opposition in the Legislative Council.[79] Consequently, the government called upon the Commonwealth to protect Victoria's Aboriginal cultural heritage. At this request, the Commonwealth used its constitutional powers under s 51(xxvi) to amend the *ATSIHP Act* in 1987 to insert part IIA, which essentially mirrored the Bill that had been proposed by Victoria.[80] This reliance by Victoria on the *ATSIHP Act* changed in 2006 with the enactment of the *Aboriginal Heritage Act 2006* (Vic) ('*AHA*'), which repealed the *AARP Act.* To be effective, it also required the Commonwealth to repeal part IIA of the *ATSIHP Act.* [81]

It is important to note at the outset that there was significant consultation (albeit flawed, according to some commentators)[82] with Indigenous Victorians prior to the enactment of the *AHA*.

Consultations with Aboriginal communities initially consisted of 'over forty "consultations" around the State prior to the drafting of the Bill, as well as 12 two-hour consultations between the Bill's release for comment and the deadline for responses two months later'.[83] Following criticisms of these initial consultations, the Minister had further meetings with 'several Indigenous and other individuals and groups, in particular the VTOLJG'.[84] This can be contrasted with the development and enactment of the *AARP Act*, for which there was no evidence of any consultation with Indigenous Victorians.[85]

There are various factors which arguably influenced the attempts by the state government to involve Indigenous Victorians in the development of the *AHA*. One important underlying factor was the growing recognition, at the international level at least, of the need to consult with Indigenous people in relation to matters

78 *ATSIHP Act* s 4.
79 See above n 14 and accompanying text.
80 Greg Gardiner, *The Aboriginal Heritage Bill 2006* (D-Brief No 2, Parliamentary Library, Parliament of Victoria, 2006) 1; Laurajane Smith, 'A History of Aboriginal Heritage Legislation in South-Eastern Australia' (2000) (50) *Australian Archaeology* 109, 115.
81 *Aboriginal and Torres Strait Islander Heritage Protection Amendment Act 2006* (Cth).
82 Brigid Anderson, Frank Hytten and Clare Land, 'Aboriginal Heritage Bill' (2006) 6 (17) *Indigenous Law Bulletin* 16; Henry Atkinson, 'Yorta Yorta Nation Aboriginal Corporation's Objection to the Cultural Heritage Bill' (2006) 6(17) *Indigenous Law Bulletin* 18.
83 Anderson, Hytten and Land, above n 82, 16.
84 Ibid 17.
85 Smith, above n 80, 114.

which affect them,[86] and the rise of the associated concept of free, prior and informed consent, which had been gaining in prominence via various declarations and treaties since the late 1980s.[87] Although matters relating to international affairs are dealt with at the Commonwealth level, the state government would have been aware of these developments. The international developments would also have given support to the expression of dissatisfaction by Victoria's Traditional Owners with the existing cultural heritage protection regime.

But probably more influential was that the political climate also favoured Indigenous involvement in the development of the *AHA*, with Labor policy (Labor being in power at the time) promoting engagement with Indigenous Victorians.[88] As Minister John Thwaites noted in his second reading speech for the Aboriginal Heritage Bill 2006 (Vic), '[w]hen the Bracks government first came to office we committed to achieving genuine partnership with indigenous people'.[89]

Does the *AHA* then embody 'genuine partnership'?

Cultural Heritage Management Plans

The purposes and objectives of the *AHA* clearly put Traditional Owners at the forefront of the protection of Aboriginal cultural heritage.[90] Thus, at the outset the *AHA* is predicated on Traditional Owners participating in the protection of their cultural heritage. Management forms an important aspect of the protection of Aboriginal cultural heritage under the *AHA*. [91]

A key feature of the *AHA* for water management is the provisions requiring approval of cultural heritage management plans ('CHMPs') by registered Aboriginal parties ('RAPs'). A RAP is an organisation which generally represents the Traditional Owners of the area for which it has been given RAP status. RAP status is determined by the Aboriginal Heritage Council,[92] which gives priority to applications made by groups who represent Traditional Owners.[93] A registered

86 See, eg, *Report of the United Nations Conference on Environment and Development, Volume I, Resolutions Adopted by the Conference*, UN Doc A/CONF.151/Rev.1 (Vol 1) (adopted 14 June 1992) annex II, ch 26.

87 See Chapter 1.

88 One of the manifestations of this policy was the 2004 amendment to Victoria's *Constitution Act 1975* (Vic) to recognise Aboriginal people as the original custodians of Victoria with 'a spiritual, social, cultural and economic relationship with their traditional lands and waters within Victoria'. *Constitution Act 1975* (Vic) s 1A(b).

89 Victoria, *Parliamentary Debates*, Legislative Assembly, 6 April 2006, 1033 (John Thwaites, Minister for Environment).

90 *AHA* ss 1(b), (c), 3(c).

91 Ibid s 3(d).

92 Ibid s 132(2)(a).

93 *AHA* s 151. See also Aboriginal Victoria, *Registered Aboriginal Parties: Guidelines for Applicants* (nd) <www.vic.gov.au/aboriginalvictoria/heritage/heritage-tools-and-publications/guides-forms-and-practice-notes-for-aboriginal-heritage-management.html#J>.

native title holder or TOGE will automatically be entitled to RAP status over the area for which it has been recognised.[94]

Various activities cannot commence unless there is an approved CHMP in place. A CHMP is evaluated and approved by the relevant RAP,[95] which may also have been involved in its development.[96] It provides recommendations for the ongoing management and protection of Aboriginal cultural heritage in the area of the activity before, during and after the activity,[97] and may also involve Traditional Owners in that ongoing management, particularly if the RAP has been involved in the development of the CHMP. Therefore, Traditional Owners will have a considerable role in the development and approval of a CHMP, and potentially a role in the subsequent management of areas for which a CHMP is in place.

The *AHA* provides that CHMPs are mandatory in certain circumstances, including if required by regulations made under the Act.[98] Under the regulations, a CHMP is required if all or part of an activity area is in an area of cultural heritage sensitivity and if all or part of the activity is a high-impact activity,[99] unless that activity is an exempt activity.[100] High-impact activities include the construction of buildings and works for specified uses, activities involving earth resource authorisations, extraction or removal of stone, timber production, and the construction of dams.[101] Exempt activities include the construction of one or two dwellings, buildings and works ancillary to a dwelling, minor works, demolition, consolidation of land, jetties, sea-bed and emergency works.[102]

Areas that have been designated as areas of cultural heritage sensitivity include waterways,[103] prior waterways,[104] ancient lakes[105] and declared Ramsar wetlands.[106] Waterway is defined quite broadly in the regulations.[107] Land within 200 metres of any of these water bodies is also an area of cultural heritage sensitivity, unless the land has been subject to significant ground disturbance.[108] Also

94 *AHA* s 151(2)–(2A).
95 Where there is no appointed RAP, a CHMP is evaluated and approved by the Secretary of the Department, or by the Aboriginal Heritage Council where the Secretary is the sponsor: ibid ss 65–66.
96 Ibid s 59(2).
97 Ibid s 42(1)(b)(ii).
98 Ibid ss 46(1)(a), 47. See also ss 48, 49, 49A, 48.
99 *Aboriginal Heritage Regulations 2018* (Vic) ('*AH Regulations*') reg 7.
100 Ibid reg 8.
101 Ibid regs 45–58.
102 Ibid regs 9–22.
103 Ibid reg 26.
104 Ibid reg 27.
105 Ibid reg 28.
106 Ibid reg 29.
107 Ibid reg 5 (definition of 'waterway').
108 Ibid regs 26(1)–(2), 27(1)–(2), 28(1)–(2), 29(1)–(2), reg 5 (definition of 'significant ground disturbance'). See also Aboriginal Victoria, *Aboriginal Heritage Act 2006 Practice Note: Significant Ground Disturbance* (nd), <www.vic.gov.au/aboriginalvictoria/heritage/heritage-tools-and-publications/guides-forms-and-practice-notes-for-aboriginal-heritage-management.html>.

designated as areas of cultural heritage sensitivity are registered cultural heritage places (which includes land within 50 metres of the place),[109] which could include water-related places.

It would therefore appear that many water sources could be the subject of a mandatory CHMP, therefore enabling participation by Traditional Owners in decisions regarding their ongoing management.

But although a CHMP provides for ongoing management of these water-related areas of cultural heritage sensitivity, its preparation and subsequent approval must first be triggered by an application to undertake an activity. To trigger a CHMP, the activity must be or involve a high-impact activity. In addition, there must not have been any significant ground disturbance in the area of cultural heritage sensitivity.[110] There is scope for the voluntary making of a CHMP,[111] but it is essentially still a reactive process given that even a voluntary CHMP will relate to a proposed activity. Further, given the costs involved with the development, approval and implementation of a CHMP, there is little incentive for a proponent to voluntarily enter into a CHMP.[112]

Thus, in areas where there are no relevant activities being undertaken or where the activity does not fall within the definition of 'high-impact activity', there is little scope for the participation of Traditional Owners in the management of water resources in that area pursuant to the *AHA*. [113]

Even if a CHMP is required, it will only relate to the area of the activity, which may not cover the entire area of the cultural heritage place. This may in effect fragment the management of the cultural heritage area where it is larger than or extends beyond the outer boundary of the activity area. Waterways are particularly susceptible to fragmentation, their length and flow potentially being the subject of a number of different activities by a number of different proponents in a number of different areas. And if an activity affects groundwater which consequentially affects a water source, but is not within the relevant area (ie 200 metres from a waterway; 50 metres from a registered cultural heritage place), then it is likely that it too may be beyond the scope of the *AHA*.

There may also be problems where a particular activity does not fit within the definition of high-impact activity and therefore does not attract the requirement for a CHMP, but where the cumulative effects on the water resource from a

109 *AH Regulations* reg 25(2).

110 This includes the 200-metre 'buffer' in relation to waterways or 50 metres for registered places.

111 *AHA* s 45.

112 At June 2011, only 108 out of a total of 1,191 CHMPs were voluntary. Department of Planning and Community Development, *Discussion Paper: Review of the Aboriginal Heritage Act 2006* (September 2011) 15. The cost of CHMPs was identified as a significant problem in the review of the *AHA*: Department of Planning and Community Development, *Summary of Submissions and Consultation – Review of the Aboriginal Heritage Act 2006* (April 2012) 16–17.

113 It is still an offence to harm Aboriginal cultural heritage (ss 27, 28), unless a cultural heritage permit is obtained (s 29).

number of similar activities by a number of different proponents, for example irrigation licences, may be quite substantial.[114]

So there are clearly limits to the role of the *AHA* in promoting the participation of Traditional Owners in the management of water resources generally.

Nonetheless, there are some benefits of the *AHA*, one being that it applies regardless of tenure. That is, Aboriginal sites located on freehold title are protected, as well as sites located on public land. In contrast, under both the *NTA* and the *TOSA*, Traditional Owners can only protect sites or places of significance on public land over which native title or Traditional Owner rights have been recognised (which will include land transferred pursuant to agreements reached under those Acts). The *AHA* thus extends that right to sites found on all land within the external boundary of an agreement/native title area. Being of state-wide application, it also applies to sites on land for which there is no *TOSA* agreement or recognition of native title.

Another benefit of the *AHA* is that one does not need to seek a declaration, nor does the site have to be registered, before it attracts the protection of the legislation. Such legislation is often referred to as providing 'blanket protection'. Blanket protection has been described as meaning

> that all areas or sites which fall within the legal definition of heritage are automatically protected by sanctions which make it an offence to cause damage or desecration to the site or area, whether or not the site has been assessed or recorded.[115]

Review of the *Aboriginal Heritage Act*

The *AHA* provided for a review of its operation after five years.[116] This review was completed in 2012. Although the streamlining of the CHMP process was noted as being one of the top issues raised in submissions to the review,[117] the resulting amendments in 2016 to the *AHA* in relation to CHMPs do not affect the provisions discussed above.

Role of the Commonwealth heritage legislation – the *Aboriginal and Torres Strait Islander Heritage Protection Act 1984* (Cth)

The *ATSIHP Act*, mentioned earlier, works in concert with other cultural heritage legislation, offering complementary protection. Others have already written about how the provisions of the *ATSIHP Act* may potentially be used to protect water-related areas of cultural heritage significance, so it is not

114 Jason Behrendt and Peter Thomson, 'The Recognition and Protection of Indigenous Interests in NSW Rivers' (2004) 3 *Journal of Indigenous Policy* 37, 112.

115 Elizabeth Evatt, 'Overview of State and Territory Aboriginal Heritage Legislation' (1998) 4(16) *Indigenous Law Bulletin* 4, 4.

116 *AHA* s 193.

117 Department of Planning and Community Development, *Summary of Submissions and Consultation – Review of the Aboriginal Heritage Act 2006* (April 2012) 12.

proposed to go into any detail here.[118] Essentially, it provides for the Commonwealth Minister to make various types of declarations to protect and preserve Aboriginal cultural heritage, on application 'made by or on behalf of an Aboriginal or group of Aboriginals'.[119] What is important to understand here is that a protection declaration made under the *ATSIHP Act* does not guarantee Indigenous access to the protected area,[120] nor does it provide for rights to manage the area. And, unlike the *AHA*, it does not provide for blanket protection – a site will only be protected if a protection declaration is obtained under the Act, which is discretionary. It does, however, contain significant penalties for contravention of a protection declaration, including a criminal conviction which can result in imprisonment.[121]

Despite such heavy penalties, or perhaps because of them, the *ATSIHP Act* has not been effective. This was made evident in the Evatt review of 1996,[122] and more recently in a 2009 discussion paper on potential reforms.[123] This discussion paper signalled the commencement of a further review of the *ATSIHP Act*. Seventy-six submissions were received by the closing date of 6 November 2009; however, no final report was produced, nor were any amendments made to the *ATSIHP Act*. In the meantime, in response to recommendation 64 of the 2011 *EPBC Act* review,[124] the Labor government did agree 'to consider incorporating the requirements of the *Aboriginal and Torres Strait Islander Heritage Protection Act 1984* into the *EPBC Act*'.[125] Since the change to a Coalition government following the federal election in November 2013, amendments to both the *ATSIHP Act* and *EPBC Act* appear to be on hold.[126]

Environmental and land management legislation

Finally, brief mention will be made of the *Heritage Rivers Act 1992* (Vic) ('*HR Act*') and the *Planning and Environment Act 1987* (Vic) ('*PE Act*') given their role in relation to water and land management. In summary, neither of these Acts

118 See, eg, Michael O'Donnell, *NAILSMA – TRaCK Project 6.2: Indigenous Rights in Water in Northern Australia* (March 2011) 127–32.

119 See, eg, *ATSIHP Act* s 9.

120 O'Donnell, above n 118, 130.

121 *ATSIHP Act* ss 22, 23.

122 Elizabeth Evatt, *Review of the Aboriginal and Torres Strait Islander Heritage Protection Act 1984* (22 August 1996).

123 Department of the Environment, Heritage, Water and the Arts (Cth), *Indigenous Heritage Law Reform – For Discussion* (August 2009) 4.

124 Allan Hawke, *The Australian Environment Act: Report of the Independent Review of the Environment Protection and Biodiversity Conservation Act 1999* (2009).

125 Australian Government, *Australian Government Response to the Report of the Independent Review of the Environment Protection and Biodiversity Conservation Act 1999* (2011) 105.

126 As at 30 May 2018.

provides any meaningful role for Indigenous Victorians to participate in water management.[127]

The *HR Act* was enacted to provide protection to various named heritage rivers and was intended to complement the *Water Act*. It was enacted just prior to the commencement of the native title era. The recognition of native title under the *NTA* over heritage rivers, however, has little, if any, practical effect on the operation of the *HR Act*. The *TOSA* specifically deals with the *HR Act*, [128] the effect being that it too has little effect on the operation of the *HR Act*. Finally, although the *HR Act*'s protection would extend to the protection of Aboriginal cultural heritage attributes of named heritage rivers, it is uncoordinated with the *AHA*.

In relation to the *PE Act*, there is little overt coordination with Victoria's water management regime or Traditional Owner interests. The *Water Act* is not explicitly referred to anywhere, there are only limited references to water,[129] and there is only one section referring to Traditional Owners (inserted pursuant to the *TOSA*) which is very limited in its application.[130] The Victoria Planning Provisions made under the *PE Act* do provide for some alignment with the *AHA*, [131] but given the deficiencies in the *AHA* noted above, this adds little to Indigenous participation in water management.

Concluding remarks

The *AHA* was a considerable improvement on the previous cultural heritage regime in Victoria. However, it suffers from a number of faults rendering it largely ineffective as a tool for Traditional Owners to participate meaningfully in the management of water resources. A key feature of the Act, CHMPs, although mandatory for waterways, are only mandatory in limited circumstances, and will only apply to the area of the activity and not the length of the waterway.

The Commonwealth's *ATSIHP Act* currently has a very limited role to play, being ineffective in both its terms and its application.

What is clear from this chapter is that there are currently two legislative regimes operating in parallel – cultural heritage protection and water management – yet they are largely uncoordinated. This exacerbates what is already a fragmented state of participation by Traditional Owners.

It is also apparent from this chapter that the same applies to Victoria's land rights and environment and land management legislative regimes. Thus, due to the lack of adequate coordination between these various regimes and the water

127 For more detail on these two Acts, see Katie O'Bryan, *From Aqua Nullius to Aqua Minimus? The Legal Recognition in Victoria of Indigenous Rights to Participate in the Management of Inland Water Resources – Lessons from Aotearoa-New Zealand* (PhD Thesis, Monash University, 2015) ch VIII.

128 *TOSA* s 26.

129 The most important of these is the definition of 'land' in s 3 which includes 'land covered with water'.

130 *PE Act* s 3 (definition of 'owner').

131 Victoria Planning Provisions (2 October 2014) cl 15.03–2.

management regime, they are inadequate guarantees of Indigenous rights to participate in water management.

This brings us to the end of the analysis of the extent to which Commonwealth and Victorian legislation recognises and facilitates Indigenous participation in water management. It is clear that although there have been some improvements in recent years, particularly at the strategy level, the complex nature of water management and the proliferation of still largely uncoordinated statutes mean that Indigenous participation in water management is highly fragmented and quite variable. It is therefore now time to turn our attention across the Tasman Sea to Aotearoa New Zealand.

Part E

Case study

Aotearoa New Zealand

8 History of water law in Aotearoa New Zealand

Introduction

The previous chapters analysed the legislative and policy recognition in Australia of Indigenous rights to participate in the management of water resources. Given the federal nature of Australia's political system and the limitations in the *Constitution* on Commonwealth power to enact water-specific legislation, the analysis focussed on Victorian legislation. Having concluded that the legislation at both levels does little to facilitate Indigenous participation in water management, this chapter examines the situation in Aotearoa New Zealand to ascertain whether its water management regime, insofar as it provides for Māori participation in the management of water resources, fares any better. Aotearoa New Zealand's similar legal and common law history and colonial context make it a useful comparator, bearing in mind some very notable differences, explored below.

The management of Aotearoa New Zealand's water resources is governed by the *Resource Management Act 1991* (NZ) ('*RMA*'), the purpose of which is 'to promote the sustainable management of natural and physical resources'.[1] The *RMA* was a major reform of Aotearoa New Zealand's resource management regime, affecting more than 50 statutes and repealing a number of major pieces of legislation,[2] including legislation relating to water. It was seen as groundbreaking at the time for combining the management of land, water and air into the one statute and was considered to be 'the largest law reform exercise in New Zealand's history'.[3] As such, there has been a wealth of material published on the *RMA* and its place in the history (and future) of Aotearoa New Zealand's environmental and resource management regime. However, the history of Aotearoa New Zealand's water management regime prior to the *RMA* is somewhat under-researched.[4]

1 *RMA* s 5(1).
2 Derek Nolan (ed), *Environmental and Resource Management Law* (LexisNexis, 5th edn, 2015) 113.
3 Ministry for the Environment, *Your Guide to the Resource Management Act* (August 2006) 5.
4 Nicola Wheen, 'A Natural Flow: The History of Water Law in New Zealand' (1997) 9(1) *Otago Law Review* 71, 72.

Accordingly, this chapter will consider, in light of our common colonial heritage, whether the development of Aotearoa New Zealand's water regulation has followed a similar trajectory to Australia and Victoria in relation to the recognition of Indigenous interests, bearing in mind that there are some significant differences between Australia and Aotearoa New Zealand that will have had an impact on its development, such as the existence of the *Treaty of Waitangi* and the Waitangi Tribunal, and Aotearoa New Zealand's political structure. The chapter then concludes with an analysis of recent developments.

Māori water management pre-colonisation

Prior to European settlement, Māori had been managing water for centuries. Some have put the arrival of Māori in Aotearoa New Zealand as early as 800 AD, others at 1200.[5] Certainly, water management by Māori predates the European-based water management regime by at least half a millennium.

Like Indigenous Australians, the Māori relationship with and world view of water is a holistic one that is embedded in the environment. Linda Te Aho expresses this Māori world view as follows:

> [A]ll are inextricably bound to the environment, especially to waterways, by virtue of whakapapa (genealogy) that derives from the creation stories of humankind in Māori cosmology. Rivers are often conceptualised as living beings, ancestors with their own mauri (life force), mana (prestige and authority) and tapu (sacredness). As such they are whole and indivisible entities, not dissected into beds, banks and waters, nor into tidal and non-tidal, navigable and non-navigable parts. The saying, 'ko au te awa, ko te awa ko au' (I am the river and the river is me), comes from the people of the Whanganui River but speaks to this connectedness that lies at the heart of the way many Māori view the world and our waterways.[6]

The Land and Water Forum ('LWF') explained the Māori relationship with water as involving three common tenets: *whakapapa*, *taonga* and *kaitiakitanga*. Those three common tenets were outlined as follows:

> a. The relationship between iwi Māori and freshwater is founded in whakapapa, which is the foundation for an inalienable relationship between *iwi* and freshwater that is recorded, celebrated and perpetuated across generations

5 Jacinta Ruru, 'Indigenous Restitution in Settling Water Claims: The Developing Cultural and Commercial Redress Opportunities in Aotearoa, New Zealand' (2013) 22 *Pacific Rim Law and Policy Journal* 311, 314, n 14. See also Philippa Mein Smith, *A Concise History of New Zealand* (Cambridge University Press, 2nd edn, 2012) ch 1; James Belich, *Making Peoples: A History of New Zealanders – From Polynesian Settlement to the End of the Nineteenth Century* (Allen Lane/Penguin Press, 1996) ch 1.

6 Linda Te Aho, 'Nga Whakataunga Waimaori – Freshwater Settlements' in Nicola Wheen and Janine Hayward (eds), *Treaty of Waitangi Settlements* (Bridget Williams Books, 2012) 102, 103.

b. Freshwater is recognised by iwi as a taonga of paramount importance
c. Kaitiakitanga is the obligation of iwi to be responsible for the wellbeing of the landscape. This obligation is inter-generational in nature. Kaitiakitanga has been given effect over the generations in many ways and differs amongst iwi and across differing circumstances.[7]

The LWF then went on further to explain:

> Waterbodies frame iwi identity – tribal traditions are transmitted across generations by continuing customary practices with waterbodies and visions for the future of iwi turn on the health and wellbeing of freshwater. The obligation to protect freshwater and to maintain and express the spiritual and ancestral relationship with freshwater so as to leave a worthy inheritance for future generations is fundamental to iwi identity.[8]

Taonga referred to by the LWF is an important concept and is specifically referred to in the Māori text of the *Treaty of Waitangi*. Translated as 'treasures' it forms part of the basis for Māori assertions of water rights that are grounded in the *Treaty of Waitangi*, to which we now turn.

Treaty of Waitangi

Signed in 1840, the *Treaty of Waitangi* is one of the most important factors to affect the recognition of Indigenous rights to manage water resources.[9] It consists of two texts, an English text and a Māori text.

The *Treaty of Waitangi* is seen as Aotearoa New Zealand's founding constitutional document, described by Sir Robin Cooke in 1990 as 'simply the most important document in New Zealand's history'.[10] Yet it is not without its controversies, notably the differences between the English and Māori texts. Much has been written about these differences,[11] so following is merely a brief summary.

7 LWF, *Report of the Land and Water Forum: A Fresh Start for Freshwater* (September 2010) 9 ('*LWF First Report*').
8 Ibid.
9 For a detailed analysis of the Treaty of Waitangi, see, eg, Matthew S R Palmer, *The Treaty of Waitangi in New Zealand's Law and Constitution* (Victoria University Press, 2008); Paul McHugh, *The Māori Magna Carta: New Zealand Law and the Treaty of Waitangi* (Oxford University Press, 1991); Claudia Orange, *The Treaty of Waitangi* (Allen & Unwin, 1987).
10 Robin Cooke, 'Introduction' (1990) *New Zealand Universities Law Review* 1. But cf Richard Boast, 'The Treaty of Waitangi and the Law' (1999) *New Zealand Law Journal* 123, 123.
11 See, eg, Bruce Biggs, 'Humpty Dumpty and the Treaty of Waitangi', in I H Kawharu (ed), *Waitangi: Māori and Pakeha Perspectives on the Treaty of Waitangi* (Oxford University Press, 1989) 300; R M Ross, 'Te Tiri o Waitangi' (1972) 6 *New Zealand Journal of History* 133; Palmer, *The Treaty of Waitangi*, above n 9, 60–63; Philip A Joseph, *Constitutional and Administrative Law in New Zealand* (Brookers, 4th edn,

The *Treaty* consists of only three articles, of which the first two are the most relevant here. In the English text, pursuant to article 1 the Māori chiefs ceded sovereignty to the Queen. But as McHugh notes, '[t]he Māori, however, had no real equivalent conception [of sovereignty]'.[12] So in the Māori text, sovereignty was translated as *kāwanatanga*, being 'a transliteration into Māori of "governor"'.[13] This term should be viewed in contrast with *rangatiratanga*, which was guaranteed to Māori in article 2 of the Māori text. *Rangatiratanga* was understood to mean 'chieftainship', which, according to Joseph, 'was a closer approximation to sovereignty than "kāwanatanga" used in art 1'.[14] However the corresponding English text guaranteed to Māori the 'full exclusive and undisturbed possession', which has quite a different emphasis.

Further, in the English text, this chieftainship, or full exclusive and undisturbed possession, was guaranteed over 'Lands and Estates Forests Fisheries and other properties'. In the Māori text, the corresponding term for 'and other properties' was '*ratou taonga katoa*'. As noted earlier, *taonga* was understood to mean 'treasures', which encompassed more than the English conception of property.[15]

But despite the differences between the English and Māori texts, and questions of sovereignty aside, what is apparent is that both the Māori concepts of *rangatiratanga* and *taonga* and the corresponding English 'translations' are a sufficient basis on which to argue that the *Treaty of Waitangi* preserves and protects Māori rights in water.

Status of the Treaty *in Aotearoa New Zealand law*

Despite being a foundational constitutional document, the *Treaty* itself is not directly enforceable by the courts unless it has been incorporated into statute, akin to the position of international treaties and conventions in both Australia and Aotearoa New Zealand.[16]

The earliest statute to incorporate the text of the *Treaty*, by way of annexure (albeit only the English version), was the *Waitangi Day Act 1960* (NZ).[17]

2014) 53–56; Te Puni Kōkiri, *He Tirohanga ō Kawa ke te Tiriti o Waitangi. A Guide to the Principles of the Treaty of Waitangi as expressed by the Courts and the Waitangi Tribunal* (Wellington, 2001) 37–41 <www.tpk.govt.nz/en/a-matou-mohiotanga/crownmaori-relations>.

12 McHugh, above n 9, 3.

13 Ibid.

14 Joseph, above n 11, 56.

15 Ibid.

16 Ibid 36; Palmer, *The Treaty of Waitangi*, above n 9, 168. *Hoani Te Heuheu Tukino v Aotea District Māori Land Board* [1941] AC 308, 324 (Viscount Simon LC); *New Zealand Māori Council v Attorney-General* ('*Lands Case*') [1987] 1 NZLR 641, 655, 667–8 (Cooke P), 691 (Somers J), 715 (Bisson J); *New Zealand Māori Council v Attorney-General* [2008] 1 NZLR 318, [62]–[76]. For a detailed discussion on the legal status of the Treaty of Waitangi in New Zealand, see, generally, Palmer, *The Treaty of Waitangi*, above n 9, ch 4.

17 Palmer, *The Treaty of Waitangi*, above n 9, 90–1.

Another Act to annex the text of the *Treaty* was the *Treaty of Waitangi Act 1975* (NZ). However, the annexing of the *Treaty* to these Acts does not give the *Treaty* the force of law generally.

These Acts, however, are unusual in annexing the full text of the *Treaty*; generally, the *Treaty* is incorporated by way of a reference to it, to its principles, or by way of a specific legislative regime.[18]

The *Treaty* is also indirectly enforceable under the principles of statutory interpretation. As Chilwell J noted in *Huakina Development Trust v Waikato Valley Authority* (coincidentally a case involving water management):

> There can be no doubt that the Treaty is part of the fabric of New Zealand society. It follows that it is part of the context in which legislation which impinges upon its principles is to be interpreted when it is proper, in accordance with the principles of statutory interpretation, to have resort to extrinsic material.[19]

The *Treaty* (and its principles) may also be indirectly enforceable as a relevant consideration under administrative law principles.[20] In the *Radio Frequencies Case*,[21] where the Court was asked whether a decision to grant radio frequency licences should be deferred until the release of a report from the Waitangi Tribunal, Richardson J noted:

> That is a conventional administrative law question and where, as here, the statute itself does not specify in so many words the criteria to be taken into account, it is a matter of determining from the scheme and purpose of the legislation what was the intention of the legislature in that regard.[22]

So despite the *Treaty* itself being a non-justiciable instrument, by virtue of the incorporation of its principles into many statutes and its role in statutory interpretation and administrative law, it still has a very significant role to play in Aotearoa New Zealand's statutory landscape.

Treaty of Waitangi Act 1975 *(NZ) and the Waitangi Tribunal*

The *Treaty of Waitangi Act 1975* (NZ) ('*TW Act*') was a milestone in Māori–Crown relations, being the culmination of decades of Māori political activism.[23] It

18 For a more nuanced analysis of how the Treaty of Waitangi may be incorporated into legislation, see generally ibid, 180–4.
19 [1987] 2 NZLR 188, 210. See also *Lands Case* [1987] 1 NZLR 641, 656 (Cooke P).
20 Joe Williams, 'Treaty Making in New Zealand' in Marcia Langton et al (eds), *Honour Among Nations? Treaties and Agreements with Indigenous People* (Melbourne University Press, 2004) 163, 165; Palmer, *The Treaty of Waitangi*, above n 9, 206–10.
21 *Attorney-General v New Zealand Māori Council* [1991] 2 NZLR 129.
22 Ibid 140. See also Casey J at 142.
23 Williams, above n 20, 165–6.

established the Waitangi Tribunal, a permanent commission of inquiry which was mandated with the task of conducting investigations into breaches of the *Treaty*. Under the *TW Act* the Tribunal was given the 'exclusive authority to determine the meaning and effect of the Treaty as embodied in the 2 texts and to decide issues raised by the differences between them'.[24]

The Tribunal began rather hesitantly,[25] hearing and reporting on only four claims prior to 1980.[26] Initially slow to be embraced by Māori,[27] the Tribunal began to take on significance following the appointment of Chief Judge Edward Taihakurei Durie as Chairperson in 1980 and the release of its Motunui-Waitara Report in 1983.[28] Notably, this Report involved the Waitara River. It has been noted that '[c]laims concerning water resources were the first to be heard by the Waitangi Tribunal'.[29]

With the expansion of the Tribunal's mandate in 1985 to include historical claims,[30] both its stature and output increased dramatically to the point where by 1990 it had been described as 'the conscience of the nation'[31] and 'the second most important institution in the country, ranking only behind parliament itself'.[32] Although in 1993 it was suggested that these pronouncements somewhat overstated the importance of the Tribunal, it was also noted that 'there is little doubt that on the whole the Tribunal commands widespread respect and legitimacy'.[33] Despite controversies over the years, this can still be said of the Tribunal today.[34]

Recommendations by the Tribunal have formed the basis for many settlements reached between Māori and the Crown, a number of which involve water rights. The Ngāi Tahu Settlement of 1998, outlined in Chapter 10, is one such settlement.

24 *TW Act* s 5(2).
25 Richard P Boast, 'The Waitangi Tribunal: "Conscience of the Nation", or Just Another Court?' (1993) *University of New South Wales Law Journal* 223, 226 (footnotes omitted).
26 Waitangi Tribunal, *Waitangi Tribunal Bibliography – Tribunal Reports, Publications and Research Reports Presented in Evidence 1975–2012 – Part 1: Waitangi Tribunal Reports, Statements and Publications* (2013) 4.
27 Andrew Sharpe, *Justice and the Māori: The Philosophy and Practice of Māori Claims in New Zealand since the 1970s* (Oxford University Press, 3rd edn, 1997) 75–6; J D Sutton 'The Treaty of Waitangi Today' (1981) *Victoria University of Wellington Law Review* 17, 34.
28 McHugh, above n 9, 311; Boast, 'Conscience of the Nation', above n 25, 226.
29 Linda Te Aho, 'Indigenous Challenges to Enhance Freshwater Governance and Management in Aotearoa New Zealand – The Waikato River Settlement' (2009) 20 *Journal of Water Law* 285, 288. See, eg, Waitangi Tribunal, *Report of the Waitangi Tribunal on the Kaituna River Claim* (1984).
30 *Treaty of Waitangi Amendment Act 1985* (NZ) s 3.
31 Paul Temm, *The Waitangi Tribunal: The Conscience of the Nation* (Random Century, 1990).
32 Boast, 'Conscience of the Nation', above n 25, 224.
33 Ibid 223.
34 Palmer suggests, however, that '[i]t would not be surprising if the Waitangi Tribunal's institutional "capital" were running out after 20 years', Palmer, *The Treaty of Waitangi*, above n 9, 198.

Finally, apart from some limited exceptions,[35] the Tribunal's recommendations are not enforceable or binding on the government or in the courts.[36] However, over the years, and particularly from the mid-1980s to the mid-1990s, the Tribunal's recommendations carried much weight and were regularly implemented.[37]

Implementation of Tribunal recommendations has fluctuated somewhat in more recent years, there being various occasions on which the government has rejected the Tribunal's recommendations.[38] However, despite (or perhaps because of) this, the Tribunal remains a fundamental feature of Aotearoa New Zealand's political landscape, and it is generally accepted that much of its early work in interpreting the meaning of the *Treaty of Waitangi* is now mainstream.[39]

In late 2012, the Tribunal published its Stage 1 Report on the National Freshwater and Geothermal Resources Claim.[40] An urgent interim version was released to inform the government on whether to proceed immediately with the sale of shares in Mighty River Power.[41] Stage 2 of the Tribunal's inquiry is still in progress. A court challenge was also instituted by Māori to prevent the sale of the shares, which was unsuccessful.[42] The court challenge, the claim and associated Tribunal reports largely relate to Māori proprietary interests in water, which is not the focus of this book. Nonetheless, it is important to be aware of these developments as they will have a significant impact on the water regulatory landscape in Aotearoa New Zealand.

Principles of the Treaty of Waitangi

The meaning of the principles of the *Treaty of Waitangi* are important, as legislation incorporating the *Treaty* will often refer to those principles. The *RMA*, analysed in more detail later, is but one example.[43] Both the Waitangi Tribunal and the courts have sought to assist in the articulation of those principles.[44] Notably, the differences between the two texts of the *Treaty* has been one of the main factors leading to the use of the term 'principles of the *Treaty of Waitangi*' in

35 Palmer, *The Treaty of Waitangi*, above n 9, 190–2, n 154.
36 *Te Rūnanga o Muriwhenua Inc v Attorney-General* [1990] 2 NZLR 641, 651 (Cooke P) But cf *Radio Frequencies Case* [1991] 2 NZLR 129, 135 (Cooke P).
37 *Lands Case* [1987] 1 NZLR 641, 661 (Cooke P). See also E Taihakurei Durie and Gordon S Orr, 'The Role of the Waitangi Tribunal and the Development of a Bicultural Jurisprudence' (1990) 14 *New Zealand Universities Law Review* 62, 62–3.
38 Palmer, *The Treaty of Waitangi*, above n 9, 110–12; 197–8.
39 Ibid 198.
40 This claim was lodged in response to the Government's proposal to privatise up to 49 per cent of four State-Owned Enterprises, including Mighty River Power.
41 Waitangi Tribunal, *Interim Report on the National Freshwater and Geothermal Resources Claim* (2012) xi.
42 *New Zealand Māori Council v Attorney-General* [2013] 3 NZLR 31.
43 *RMA* s 8. See, eg, *Conservation Act 1987* (NZ) s 4; *Crown Pastoral Land Act 1998* (NZ) s 25; *State Owned Enterprises Act 1986* (NZ) s 9.
44 For a detailed discussion, see generally Te Puni Kōkiri, above n 11.

statutory *Treaty* clauses rather than the use of 'terms or provisions of the *Treaty*'.[45] McKay J, in the *Broadcasting Case*, explained it thus:

> It is the principles of the Treaty which are to be applied, not the literal words. The English and Māori texts in the first schedule to the Treaty of Waitangi Act 1975 are not translations the one of the other, and the differences between the texts and shades of meaning are less important than the spirit.[46]

The other main factor has been the need for the *Treaty* to be able to adapt to changing circumstances. The words of Cooke P here are instructive: 'The Treaty obligations are ongoing. They will evolve from generation to generation as conditions change.'[47]

Despite various attempts to distil a list of principles,[48] there is no definitive statement of the content of the *Treaty* principles.[49] However, there are some more settled principles and common themes that can be extracted from these sources, of which the following are the most relevant.

The first is the principle of partnership, which includes an obligation to act honourably, reasonably and in good faith. According to Sir Robin Cooke, writing extra-curially in 1990, this did not necessarily mean equal shares in the partnership, but one that connoted 'a continuing relationship between parties working together and owing each other duties of reasonable conduct and good faith'.[50]

Another is the principle, or duty, of active protection, by which the Crown is obliged to take positive steps to protect Māori interests. This was recognised by Cooke P in the *Lands Case* where he stated that 'the duty of the Crown is not merely passive but extends to active protection of Māori people in the use of their lands and waters to the fullest extent practicable'.[51]

Although arguably not a *Treaty* principle in its own right,[52] the duty to consult is another common theme that has emerged from the sources, and is of particular relevance to the *RMA* and the underlying theme of this book.

Therefore, by virtue of the existence of the *Treaty* and the incorporation of its principles into various statutes, Māori have a much stronger moral and, in some

45 Ibid 74.
46 *New Zealand Māori Council v Attorney-General* [1992] 2 NZLR 576, 590. McKay J was here paraphrasing Cooke P in the *Lands Case* [1987] 1 NZLR 641, 661–3.
47 *Te Rūnanga o Muriwhenua v Attorney-General* [1990] 2 NZLR 641, 656. See also Waitangi Tribunal, *Muriwhenua Report* (1997) 386.
48 Janine Hayward, 'Appendix – The Principles of the Treaty of Waitangi', in Volume 2 of Alan Ward, *National Overview of the Waitangi Tribunal Rangahaua Whanui Series* (GP Publications, 1997); New Zealand Law Commission, *Māori Custom and Values in New Zealand Law* (Study Paper 9, 2001) [338]–[351].
49 *Carter Holt Harvey Ltd v Te Rūnanga o Tūwharetoa Ki Kawerau* [2003] 2 NZLR 349, [27] (Heath J).
50 Cooke, above n 10, 5.
51 *Lands Case* [1987] 1 NZLR 641, 664.
52 Ibid 693 (Lord Somers).

instances, legal basis than Indigenous Australians on which to argue that they are not just another stakeholder but should have an influential seat at the decision-making table.

Treaty *settlements*

The *Treaty* claims settlement process has also had a significant impact on Māori participation in water management, and an analysis of various settlements involving water management rights is the subject of Chapter 10. Since its inception, the Waitangi Tribunal has played a major role in the settlement of *Treaty* claims, its reports and associated recommendations being an important source used to inform negotiations. However, Māori can also directly negotiate a settlement with the Crown, the Waikato River settlement of 2010 being an example of a settlement achieved by way of direct negotiations.

The preceding sections have briefly outlined the history and status of the *Treaty of Waitangi* in Aotearoa New Zealand, touching in a general way on how it has facilitated Māori participation in water management. The history of Aotearoa New Zealand's water management regime is the focus of the following section, highlighting various other factors that may also have had an impact, with comment on any relevant comparisons that can be made with Australia and Victoria.

The development of Aotearoa New Zealand's water management regime

A brief history

Following colonisation in 1840,[53] Aotearoa New Zealand's system of water regulation, like Victoria's, was initially based on the English common law doctrine of riparian rights. And, as in Victoria, the common law doctrine was eventually superseded in a rather ad hoc fashion by a statutory regime.[54]

The statutes of note during this transition from common law to statutory-based rules included the *Soil Conservation and Rivers Control Act 1941* (NZ) ('*SCRC Act*') and the *Water and Soil Conservation Act 1967* (NZ) ('*WSC Act*'). The *SCRC Act* was notable because it provided for a more coherent management structure for water, insofar as it related to land drainage and flood control, floods being a major issue at that time. It achieved this with the establishment of catchment boards based on water catchment districts, and of the Soil Conservation and Rivers Control Council to oversee them. The *WSC Act* further enhanced the

53 Joseph, above n 11, 109–10; Peter Spiller, Jeremy Finn and Richard Boast, *A New Zealand Legal History* (Brookers, 2nd edn, 2001) 65.

54 B H Davis, 'New Control Over Natural Water' (1968) *New Zealand Law Journal* 105, 105. See also Wheen, 'A Natural Flow', above n 4, for further information on the legislative developments.

management structures but, more importantly, by virtue of s 21, extinguished most, if not all, common law riparian rights.[55]

Section 21 of the *WSC Act* bears some similarity with s 4 of the *Irrigation Act 1886* (Vic) in that various activities were vested in the Crown. Both the *WSC Act* and the *Irrigation Act* vested the use of water in the Crown. In Victoria, this was extended in 1905 to include flow and control, which has been preserved in s 7(1) of the *Water Act 1989* (Vic). However, a point of difference was that s 21 of the *WCS Act*, unlike the Victorian statutes, did not specifically mention the vesting of control or management of waters.

Described in 1968 as 'the pinnacle of the water control pyramid',[56] the *WSC Act* was enacted during a time when Aotearoa New Zealand was in the process of constructing numerous dams, a process which commenced somewhat later than in Australia, and particularly in Victoria, where the construction of dams commenced in the late 1800s (although it continued well into the same timeframe). But the purposes for which the dams were being built were similar; both Australia and Aotearoa New Zealand were building dams for irrigation and hydro-electricity. Water quality issues were also emerging as an issue in Aotearoa New Zealand during this time, which led in 1971 to catchment authorities being given powers over water quality.[57]

The rate of dam construction in the 1950s and 1960s and the potential for construction to continue became of concern to many New Zealanders in the 1970s, leading to efforts to protect icon rivers and lakes from development. This resulted in amendments to the *WSC Act* in 1981.[58] The 1970s and the decades leading up to it were also a period of Māori political activism, influenced to a large extent (as was the case in Australia) by the civil rights movement in the United States and other international developments in human/Indigenous rights.[59] It was therefore a period in which Māori aspirations became much more of a national focus, and which led to the establishment in 1975 of the Waitangi Tribunal. These Māori political developments were not, however, reflected in the 1981 amendments to the *WSC Act*. A similar disconnect was also shown to be evident in the development of Victoria's water management regime.[60]

The focus in the 1980s then turned to ecosystems, for which the medley of existing legislation was found to be inadequate. This led to the development and subsequent enactment of the *RMA*. As noted earlier, it was the period when the Waitangi Tribunal began to take a much more prominent place in Aotearoa New

55 *LWF First Report*, above n 7, 9. See also Davis, above n 54; *Glenmark Homestead Ltd v North Canterbury Catchment Council* [1978] 1 NZLR 407, 413 (Woodhouse J).

56 Davis, above n 54, 105.

57 Eric Pyle et al, 'Establishing Water Management in Law: The New Zealand Experience' (2001) 37(4) *Journal of the American Water Resources Association* 783, 785. This was pursuant to the *Water and Soil Conservation Amendment Act (No 2)* 1971 (NZ).

58 *Water and Soil Conservation Amendment Act 1981* (NZ).

59 Mein Smith, above n 5, 238.

60 See Chapter 3.

Zealand's political landscape. It was also a decade of judicial activism in which the courts handed down a number of important cases relating to Māori rights, the most significant of these being the 1987 *Lands Case.* [61]

Current water management regime: the Resource Management Act

The *RMA* was revolutionary in that it combined the management of land, water and air in the one statute. It repealed a large number of Acts relating to resource management, including the *WSC Act.* [62] However, in terms of the provisions relating to the management of water, it has been argued that these were not revolutionary at all, but simply a 'natural progression' of existing arrangements.[63]

Some of the existing arrangements that continued under the *RMA* include the preservation of the Crown's rights that were contained in s 21 of the *WSC Act,* [64] and the preservation of statutory-based riparian rights to take and use water for domestic and stock purposes, but with the added proviso that the taking and use does not have or is not likely to have an adverse impact on the environment.[65]

There was, however, considerable involvement of Māori in its development, resulting in the recognition of Māori interests, such as s 8. Section 8, which requires decision-makers under the Act to take into account the principles of the *Treaty of Waitangi*, has been a large hook on which Māori have hung their aspirations for resource management. It has been suggested that '[t]he law reform consultation exercise was the largest and most comprehensive process for participation by Māori in the formulation of policy and law ever undertaken by the New Zealand Government'.[66] Others have written in more detail about Māori participation in the development of the *RMA*;[67] however, this is a major point of difference between the development of the *RMA* and the development of Victoria's *Water Act* (enacted a mere two years earlier), for which, despite a comprehensive consultation process, there is no evidence of consultations specifically targeting Aboriginal Victorians.

But not all the credit for the inclusion of Māori interests in the *RMA* can be attributed to participation in the consultation process for its development. The development of legislation is necessarily part of a political process, and there are

61 [1987] 1 NZLR 641. Also of some significance was *Te Weehi v Regional Fisheries Officer* [1986] 1 NZLR 680 and *Huakina Development Trust v Waikato Valley Authority* [1987] 2 NZLR 188.

62 For a full list of repealed Acts, see *RMA* sch 6.

63 Wheen, 'A Natural Flow', above n 4, 71.

64 *RMA* s 354(1)(b).

65 Ibid s 14(3)(b).

66 Diane L Crengle, 'Perspectives on Māori Participation Under the Resource Management Act', in Inter-Commission Task Force on Indigenous Peoples, *Indigenous Peoples and Sustainability – Cases and Actions* (International Books, 1997) 344.

67 Janet Stephenson, 'Recognising Rangatiratanga in Resource Management for Māori Land: A Need for a New Set of Arrangements' (2001) 5 *New Zealand Journal of Environmental Law* 159.

some important differences as well as some similarities between Aotearoa New Zealand and Victoria/Australia that deserve mention.

System of government

Aotearoa New Zealand has a national unicameral system in which 'Parliament is supreme and has no formal limits to its law making powers'.[68] It is a two-tiered system, with some form of provincial or local government having existed since the very early days of the colony.[69] The place of local government is not based on any 'constitutional entitlement',[70] but derives its powers from legislation enacted by the national government.[71]

To a large extent, the power of the Aotearoa New Zealand Parliament resembles the power of state parliaments in Australia (and Queensland in particular, which also has a unicameral parliament), which generally speaking have plenary power to legislate.[72] Aotearoa New Zealand has a *Constitution Act 1986* (NZ) and a *Bill of Rights Act 1990* (NZ),[73] but like the equivalent *Constitution Act 1975* (Vic) and the *Charter of Human Rights and Responsibilities Act 2006* (Vic), the Aotearoa New Zealand statutes can be amended by ordinary legislation subject to manner and form requirements prescribed by statute.[74] The status of local government in Australia is also similar to Aotearoa New Zealand in that its existence depends on state legislation; in Victoria, this is the *Local Government Act 1989* (Vic). These similarities make comparisons between Aotearoa New Zealand and a particular Australian state (in this case Victoria) particularly apt in the context of this book.

Māori representation in Parliament and voting

Unlike Australia, Aotearoa New Zealand's political system provides for Māori seats in Parliament, a feature first introduced in 1867 on a temporary basis but made permanent in 1876.[75] There were initially four Māori electorates, but following

68 Jacinta Ruru, 'Undefined and Unresolved: Exploring Indigenous Rights in Aotearoa New Zealand's Freshwater Legal Regime' (2009) 20 *Journal of Water Law* 236, 237.

69 Geoffrey Palmer and Matthew Palmer, *Bridled Power: New Zealand's Constitution and Government* (Oxford University Press, 4th edn, 2004) 247.

70 Kenneth A Palmer, *Local Government Law in New Zealand* (Law Book, 2nd edn, 1993) 23.

71 *Local Government Act 2002* (NZ).

72 *Union Steamship Co v King* (1988) 166 CLR 1. There are some limitations on the states' power to legislate imposed by the *Commonwealth Constitution*, eg ss 52, 90 and 109.

73 There are numerous other Acts which form the source of governmental power; see Joseph, above n 11, 28–9. See also Geoffrey Palmer and Matthew Palmer, *Bridled Power*, above n 69, 5.

74 For a detailed discussion on manner and form in Aotearoa New Zealand, see Joseph, above n 11, ch 16.

75 Electoral Commission (NZ), *Māori Representation* (8 February 2013) <www.elections.org.nz/māori-and-vote/māori-representation>.

the introduction of the mixed-member proportional ('MMP') system in 1993, that number has risen to seven.[76] So from very early on, Māori have had a permanent, statutory-based voice in Parliament.

It has also been noted that the introduction of the MMP voting system in 1993 has put a brake on Parliament's ability to pass legislation as it 'makes it much less likely that any one political party will get a majority of seats in Parliament'.[77] Consequently, several parties will need to form a government either by way of a coalition government or by one party forming a minority government with the support of another party or parties. This has, to some extent, been to the benefit of Māori, who have been able to have a much stronger voice in Parliament in recent years.[78]

The power of the ballot box is another way in which Indigenous people can influence the development of legislation, including water management legislation, and therefore the right to vote is important. Māori, with around 17.5 per cent of Aotearoa New Zealand's population,[79] have a much greater ability to be a significant voting force than Indigenous Victorians, who make up only 0.7 per cent of the Victorian population.[80] However, as voting is not compulsory in Aotearoa New Zealand (although it is a legal requirement to register to vote),[81] Māori must necessarily be convinced that it is in their best interests to vote, as must all New Zealanders. But even accounting for the non-compulsory nature of Aotearoa New Zealand's voting system and the lower turnout of Māori voters compared with the rest of the population,[82] the Māori vote would still be a much more influential one compared with that of Indigenous Victorians.

So it is arguably a combination of all of the above political factors which, first, led to the recognition of the need to consult with Māori in the development of the *RMA*, and, second, meant that those consultations were taken seriously, the outcome being the recognition of Māori interests in the *RMA*.

Yet this does not appear to have been the case with the reform of local government, which occurred concurrently with, and was necessary for, the successful

76 Electoral Commission (NZ), *Māori Representation* (22 March 2018) <www.elections.org.nz/voting-system/māori-representation>.

77 Geoffrey Palmer and Matthew Palmer, *Bridled Power*, above n 69, 14.

78 Ann Sullivan, 'Tōrangapū – Māori and political parties – National, New Zealand First, Māori and Mana parties', *Te Ara: The Encyclopedia of New Zealand* (18 July 2016) <www.TeAra.govt.nz/en/torangapu-māori-and-political-parties/page-4>.

79 Statistics New Zealand, *2013 Census Quick Stats About Māori* (2013) 6 <http://archive.stats.govt.nz/Census/2013-census/profile-and-summary-reports/quickstats-about-māori-english.aspx>.

80 Australian Bureau of Statistics, *2076.0 – Census of Population and Housing: Characteristics of Aboriginal and Torres Strait Islander Australians, 2011 – Population Distribution and Structure* (16 February 2018) <www.abs.gov.au/ausstats/abs@.nsf/Lookup/2076.0main+features1102011>.

81 Palmer and Palmer, *Bridled Power*, above n 69, 23.

82 Eljon Fitzgerald, Brendan Stevenson and Jacob Tapiata, *Māori Electoral Participation – A Report Produced for the Electoral Commission* (Massey University, May 2007) 6–7.

implementation of the resource management law reforms ('RMLR') in the *RMA*. In discussing the close relationship between local government reform and the RMLR in the context of the *Treaty of Waitangi*, it was noted by Dixon and Wrathall in 1990 that '[u]nfortunately the extensive consultation process undertaken by Ministry for the Environment staff during RMLR has not been present in Local Government Reform'.[83]

One might wonder why, given what has been outlined above. The political machinations at the time are too intricate to be discussed here, but one of the major factors that arguably distinguished the local government reform process from the RMLR process was the fact that statutes in relation to the environment and conservation already made reference to the principles of the *Treaty of Waitangi.* [84] In addition, recognition of Māori interests had been part of resource management since the enactment of the *Town and Country Planning Act 1977* (NZ) (repealed by the *RMA*), which 'introduced the first recognition of Māori interests within planning law'.[85] Thus, it has been argued, it would have been much more difficult to ignore Māori interests in the RMLR process.[86] This supports the importance of having legislative recognition of Indigenous interests to ensure that Indigenous voices are heard.

Local government reform

This reform of local government was part of an overall programme of economic, social and administrative reform being undertaken by the New Zealand government.[87] The reforms involved, among other things, a streamlining of central government agencies and a reduction of local government authorities from approximately 800 to 88.[88]

As part of this process, many agencies were privatised, their functions being transferred to State-Owned Enterprises. Responsibility for environmental management was vested in three new agencies: the Department of Conservation, the

83 Jennifer Dixon and Aileen Wrathall, 'The Reorganisation of Local Government: Reform or Rhetoric?' (1990) *New Zealand Journal of Geography* 2, 3. See also Crengle, above n 66, 345–6.
84 *Conservation Act 1987* (NZ) s 4, *Environment Act 1986* (NZ) preamble.
85 Stephenson, above n 67, 176.
86 Elizabeth Jane Kelsey, *Rogernomics and the Treaty of Waitangi* (PhD Thesis, University of Auckland, 1991) 468.
87 For more information on the local government reforms, see Kenneth A Palmer, above n 70, and, more recently, Kenneth A Palmer, *Local Authorities Law in New Zealand* (Brookers, 2012).
88 L Gow, 'New Zealand Resource Management Act: Implementing A Major Planning Law Reform' in Donna G Craig, Nicholas A Robinson and Koh Kheng-Lian (eds), *Capacity Building for Environmental Law in the Asian and Pacific Region: Approaches and Resources* (Asian Development Bank, 2002) 431, 431. But cf Owen Furuseth and Chris Cocklin, 'An Institutional Framework for Sustainable Resource Management: The New Zealand Model' (1995) 35 *Natural Resources Journal* 243, 256.

Ministry for the Environment and the Parliamentary Commissioner for the Environment.[89]

The reforms to local government created a two-tier framework comprised of regional councils and territorial authorities.[90] The boundaries of regional councils were, to a considerable extent, based on catchment boundaries.[91]

An important element of these reforms was the devolution of many powers from central to local government, with central government maintaining oversight. A consequence of this, however, was to 'confer substantial powers on bodies which bore no obligation to comply with the Treaty'.[92] This problem was alleviated to some extent in the *RMA* by the inclusion of s 8, which requires all decision-makers under the *RMA*, including local government authorities, to take into account the principles of the *Treaty of Waitangi*. [93] More recent local government legislation has also lessened this problem.[94]

The enactment of the *RMA* established that water allocation and management was clearly the responsibility of regional councils, created pursuant to the local government reforms noted above.[95] This was an extension of the role previously undertaken by catchment authorities which had been established and developed under previous legislation.[96] Under the *RMA*, the central government's role consisted of setting national environmental standards and national policy statements,[97] a role it chose not to exercise for some years in relation to water management.[98] However, this has changed in more recent years.[99]

Recent national developments in water management

From the Sustainable Water Program of Action 2003 to the National Policy Statement for Freshwater Management 2017

Aotearoa New Zealand's central government started to take a more prominent role in 2003 when it commenced its Sustainable Water Programme of Action, one of the four priority areas in the government's Sustainable Development

89 Furuseth and Cocklin, above n 88, 253.
90 Ibid 256.
91 Nicola Wheen, 'An Updated History of New Zealand Environmental Law' in Eric Pawson and Tom Brooking (eds), *Making a New Land: Environmental Histories of New Zealand* (Otago University Press, revised edn, 2013) 277, 286.
92 Kelsey, above n 86, 468.
93 *RMA* s 8.
94 *Local Government Act 2002* (NZ) s 4.
95 *RMA* s 30.
96 *Soil Conservation and Rivers Control Act 1941* (NZ); *Water and Soil Conservation Act 1967* (NZ).
97 *RMA* ss 43AA–46B.
98 Pyle et al, above n 57, 787.
99 This is in relation to the *RMA* as a whole, and arguably started with the 2005 amendments to the *RMA*: New Zealand, *Parliamentary Debates*, 2 August 2005, 22273 (David Benson-Pope, Associate Minister for the Environment).

Programme of Action. A significant impetus behind the Sustainable Development Programme of Action and Sustainable Water Programme of Action was the World Summit for Sustainable Development held in Johannesburg in 2002, in which water was an important issue.[100] Following the release of a discussion paper in December 2004,[101] extensive consultations were undertaken, including with Māori.[102] These consultations resulted in the publication of a number of reports,[103] the report on the Māori consultations being by far the most extensive.[104] An implementation package was released in April 2006. In November that year, the Minister for the Environment then decided 'that it would be desirable to issue a national policy statement on freshwater management',[105] pursuant to the processes in the *RMA*. Following consultations with *iwi* and other stakeholders,[106] the Minister prepared a draft *National Policy Statement for Freshwater Management* ('*NPS-FM*') and appointed a Board of Inquiry to consider the draft, the report of which was released in January 2010.[107] In the meantime, the government announced a new strategy for water: New Start for Fresh Water. As part of this new strategy, the government asked the Land and Water Forum ('LWF'), a body comprising 'a range of industry groups, environmental and recreational NGOs, *iwi*, scientists, and other organisations with a stake in freshwater and land management'[108] to consider reform of New Zealand's freshwater management system, including its views on the proposed *NPS-FM*. The LWF subsequently produced three reports[109] which led to the 'Fresh Start for Fresh Water' reforms, including the *NPS-FM*, approved in May 2011.

100 Department of Prime Minister and Cabinet (NZ), *Sustainable Development for New Zealand, Program of Action* (January 2003) 6.
101 Ministry for the Environment (NZ), *Freshwater for a Sustainable Future: Issues and Options – A Public Discussion Paper on the Management of New Zealand's Freshwater Resources* (December 2004).
102 Seventeen *hui* were held around Aotearoa New Zealand, along with 19 general public meetings; 292 written submissions were received, including approximately 28 from Māori: Ministry for the Environment, *Testing the Water – Report on the Sustainable Water Program of Action Written Submissions* (July 2005) 4.
103 Ministry for the Environment (NZ), *Wai Ora: Report of the Sustainable Water Programme of Action Consultation Hui* (July 2005) ('*Wai Ora Report*'); Ministry for the Environment (NZ), *Report of the Sustainable Water Program of Action Public Meetings* (July 2005); Ministry for the Environment (NZ), *Report of the Sustainable Water Program of Action Local Government Workshops* (July 2005).
104 *Wai Ora Report*, above n 103.
105 David Sheppard et al, *Report and Recommendations of the Board of Enquiry into the Proposed National Policy Statement for Freshwater Management* (2010) 7.
106 Invitations to comment were sent to 300 *iwi* organisations and other stakeholders: ibid.
107 Ibid.
108 Land and Water Forum <www.landandwater.org.nz>.
109 *LWF First Report*, above n 7; LWF, *Second Report of the Land and Water Forum: Setting Limits for Water Quality and Quantity – Freshwater Policy and Plan-Making through Collaboration* (April 2012); LWF, *Third Report of the Land and Water Forum: Managing Water Quality and Allocating Water* (October 2012).

Changes to the *NPS-FM* were then proposed in November 2013.[110] Twenty-six public consultation meetings, of which more than half were *hui*, were held to discuss the proposed changes.[111] Submissions were then received on the proposed changes, including 35 from Māori/*Iwi*, [112] and the amendments to the *NPS-FM* were approved in July 2014. The *NPS-FM* was amended again in 2017, following further public consultations and submissions, of which 21 were from Māori/*Iwi*. [113] Also involved in the development of the proposed amendments was the Iwi Advisors Group.[114]

The discussion above indicates that Māori had substantial input into water reform and the development of the *NPS-FM* including the subsequent amendments, the result of that input being substantial recognition in the *NPS-FM* of Māori interests in freshwater management.[115] For example, the *NPS-FM* recognises *Te Mana o te Wai* (the integrated and holistic wellbeing of a freshwater body) as being of national significance and integral to the management of freshwater.[116]

There is little doubt that Māori interests in freshwater management are an essential component of the *NPS-FM*, resulting from significant Māori input into its development. That input can be attributed to various factors, including the legislative requirement in the *RMA* (under which the *NPS-FM* is made) to consult with *iwi*, [117] as well as, more generally, a permanent Māori voice in Parliament and a proportionally larger Indigenous population, as discussed earlier.

Freshwater Reform 2013 and beyond

The changes to the *NPS-FM* form part of the Aotearoa New Zealand government's wider reform package. This reform package commenced in March 2013 with the release of a discussion paper, *Freshwater Reform 2013 and Beyond*. [118] The purpose of the discussion paper was to set out and receive feedback from the public on 'the Government's approach to reforming New Zealand's freshwater

110 Ministry for the Environment, *Proposed Amendments to the National Policy Statement for Freshwater Management 2011: A Discussion Document* (November 2013).

111 Of the 26 meetings held, 12 were public meetings and 14 were hui: Ministry for the Environment, *National Policy Statement for Freshwater Management Amendment Consultation 2013* (13 February 2014) <www.consultation.mfe.govt.nz/content/national-policy-statement-freshwater-management-amendment-consultation-2013> (NB: copy of webpage on file with the author).

112 Ministry for the Environment, *Report and Recommendations on the Proposed Amendments to the National Policy Statement for Freshwater Management and Public Submissions* (July 2014) 8.

113 Ministry for the Environment, *Submissions Report and Recommendations on Proposed Amendments to the National Policy Statement for Freshwater Management 2014* (2017) 6.

114 Ibid.

115 See, eg, Preamble, Part AA, Part D, Appendix 1.

116 *NPS-FM* 7.

117 *RMA* ss 46–53.

118 Ministry for the Environment, *Freshwater Reform 2013 and Beyond* (March 2013).

management system'[119] in the face of mounting pressure on Aotearoa New Zealand's water resources. Following the release of the discussion paper, around 2,000 people attended over 50 public meetings held in March 2013, as well as *hui*, council and stakeholder meetings.[120] Comments on the discussion paper closed on 8 April 2013. Despite the relatively short timeframe from the release of the discussion paper to the submission closing date, 368 written submissions were received, of which 36 came from *iwi, hapū* and *iwi*/Māori organisations.[121] One of the key concerns raised by *iwi*/Māori submitters was the need for 'a clear role for iwi/Māori in decision-making that is more than advisory'.[122]

The next step was the release in February 2016, of a consultation paper, *Next Steps for Freshwater*, the feedback from which informed the 2017 amendments to the *NPS-FM*. [123]

This wider reform has also resulted in amendments to the *RMA*. [124]

Concluding remarks

Māori have had a much stronger voice in the development of Aotearoa New Zealand's water management regime compared with Indigenous people in Victoria's (and Australia's) water management regime. It is also evident in the most recent proposals that they are continuing to have such a voice. This has been attributed to various factors, the *Treaty of Waitangi* and the Waitangi Tribunal being of particular influence in that regard. But other factors, such as the greater power of the Māori vote and Aotearoa New Zealand's political structure, have also played an important role.

As part of this analysis, two of the main avenues for Māori participation in water management have been identified, namely the *RMA* and Treaty settlements. There is, however, a third avenue which needs to be considered if one wants to undertake a comprehensive comparison with Australia, that avenue being native title. The next two chapters will therefore look in more detail at all three avenues, commencing with native title.

119 Ibid 7.

120 These meetings related to both the freshwater reforms and the wider resource management reform proposals.

121 Ministry for the Environment, *Freshwater Reform and Beyond: Summary of Feedback* (10 July 2013) <www.mfe.govt.nz/issues/water/freshwater/freshwater-reform-2013/summary.html> (NB: copy of webpage on file with the author).

122 Ibid.

123 Ministry for the Environment, *Submissions Report and Recommendations on Proposed Amendments to the National Policy Statement for Freshwater Management 2014* (2017) 5.

124 *Resource Legislation Amendment Act 2017* (NZ).

9 Engaging with Māori rights – native title, and the *Resource Management Act 1991* (NZ)

Māori water rights and native title

Given the extensive analysis in Chapter 4 of native title in Australia as a vehicle for Indigenous participation in water management, it is important for comparative purposes to consider what role, if any, native title has played in Aotearoa New Zealand.[1]

Native title[2] in Aotearoa New Zealand has had a somewhat different trajectory compared with Australia. Both have a common starting point, that being the role of English common law upon the acquisition of sovereignty. As in Australia, upon the acquisition of sovereignty '[t]he laws of England were applied in New Zealand "so far as applicable to the circumstances thereof"'.[3] But that is where the two countries diverge, both in how sovereignty was acquired and the extent to which the common law was adapted to suit local circumstances. Australia was originally seen as 'desert and uncultivated' in common law terms[4] which allowed for acquisition by settlement, and with little adaptation of the common law to local circumstances.[5]

Aotearoa New Zealand, on the other hand, 'was acquired in haphazard fashion, with little regard for constitutional forms'.[6] Although a number of grounds have been identified as the basis for the acquisition of Aotearoa New Zealand,[7] the British government did make two formal proclamations of sovereignty: a proclamation for the North Island based on cession occurring by virtue of the *Treaty of Waitangi*, and a proclamation for the South and Stewart Islands based on discovery.[8] These differing circumstances and their effect on the reception of British

1 For a detailed comparative analysis of native title in Australia and New Zealand, see Simon Young, *The Trouble with Tradition: Native Title and Cultural Change* (Federation Press, 2008).
2 Also referred to as Māori, Aboriginal or customary title.
3 *Attorney-General v Ngati Apa* [2003] 3 NZLR 643, 652 (Elias CJ).
4 The equivalent of the international law term *terra nullius.*
5 See Chapter 2.
6 Philip A Joseph, *Constitutional and Administrative Law in New Zealand* (Brookers, 4th edn, 2014) 41.
7 Ibid 47.
8 Ibid 46.

common law into Aotearoa New Zealand contributed to a different approach to native title.

Australia did not recognise the existence of native title until the 1992 High Court decision in *Mabo.* [9] Aotearoa New Zealand, by contrast, recognised native title from the outset, both in statute and by the courts.[10] The 1847 case of *R v Symonds* [11] is held up as representing 'the foundational principles of the common law relating to Māori'.[12] In this case, the Supreme Court (precursor to the now High Court)[13] noted the existence of native title, and articulated the role of the *Treaty of Waitangi* in preserving it.[14]

In 1862, New Zealand's colonial government commenced a process of statutory conversion of native title to freehold ownership under the *Native Land Act 1862* (NZ). That Act established a Native Land Court to determine title to land held communally by Māori, and to convert that title to something resembling freehold title under British law.[15] Thus began a process of extinguishment of native title over land. The nadir of mediated dispossession was contained in ss 84–86 of the *Native Land Act 1909* (NZ) which ended customary claims to land and validated all alienations of un-surrendered land. So '[b]y 1911 the area of dry land not yet investigated by the Native Land Court, with the consequent extinguishment of customary title, was tiny'.[16] Extinguishment of native title over the beds of navigable rivers was purportedly effected by s 14 of the *Coal-mines Amendment Act 1903* (NZ),[17] although it has been noted 'that the legal situation vis-à-vis inland waterways in New Zealand – especially as to their ownership – is at best indeterminate'.[18]

9 (1992) 175 CLR 1.
10 Joseph, above n 6, 97.
11 [1847] NZPCC 387.
12 Robert J Miller et al, *Discovering Indigenous Lands: The Doctrine of Discovery in the English Colonies* (Oxford University Press, 2010) 215; Mark Hickford, 'Settling Some Very Important Principles of Colonial Law: Three "Forgotten" Cases of the 1840s' (2004) 35 *Victoria University of Wellington Law Review* 1, 1–2.
13 The Supreme Court was established in 1841 and was reconstituted in 1979 as the High Court.
14 *R v Symonds* [1847] NZPCC 387, 390 (Chapman J).
15 For a detailed discussion on the *Native Land Act 1862* (NZ), its later variations and the Native Land Court, see Richard Boast, *Buying the Land, Selling the Land: Governments and Māori Land in the North Island 1865–1921* (Victoria University Press, 2008) ch 2.
16 David V Williams, 'Wi Parata is Dead, Long Live Wi Parata' in Claire Charters and Andrew Erueti (eds), *Māori Property Rights and the Foreshore and Seabed: The Last Frontier* (Victoria University Press, 2007).
17 Waitangi Tribunal, *Whanganui River Report* (1999) 268. But cf *Te Runanganui o Te Ika Whenua Inc Society v Attorney-General* [1994] 2 NZLR 20, 26 (Cooke P). See also F M (Jock) Brookfield, 'The Waitangi Tribunal and the Whanganui River-Bed' [2000] *New Zealand Law Review* 1.
18 Ben White, *Inland Waterways: Lakes* (National Theme Q, Waitangi Tribunal Rangahua Whanui Series, March 1998) 6.

Despite *R v Symonds* and various statutes which recognised customary interests in land, in the infamous 1877 case of *Wi Parata v Bishop of Wellington* ('*Wi Parata*'),[19] the Supreme Court held that Māori did not exist as sovereign peoples with native title rights capable of being enforced in the courts. Māori claims based on native title were held by the court to be non-justiciable, a reason being that Māori were primitive and had no system of governance. Prendergast CJ delivered the judgment for the Supreme Court, stating that:

> On the foundation of this colony, the aborigines were found without any kind of civil government, or any settled system of law. … The Māori tribes were incapable of performing the duties, and therefore assuming the rights of a civilized community.[20]

He went on famously to say that the *Treaty of Waitangi* 'must be regarded as a simple nullity'.[21]

Condemnation was thrown on this decision by the Privy Council in *Nireaha Tamaki v Baker* [22] in which Lord Davey stated that the decision in *Wi Parata* 'goes too far' and that it was 'rather late in the day' for such arguments (denying the existence of customary law) to be put before New Zealand courts.[23]

This, however, did not spell the end of *Wi Parata*. The New Zealand Court of Appeal was so incensed by the Privy Council's condemnation that, in an unprecedented reaction, it issued a formal protest.[24] Cases in the early twentieth century continued to follow or cite *Wi Parata* with approval, ignoring the condemnation of it by the Privy Council.[25] This continued into the 1960s, with *Re the Bed of the Wanganui River* [26] and *Re the Ninety-Mile Beach* ('*Ninety-Mile Beach*').[27] In *Ninety-Mile Beach* the Court of Appeal held that Māori title in the foreshore and seabed had been extinguished.

However, due to a 'noticeable political and legal shift'[28] in the 1970s, including the establishment of the Waitangi Tribunal in 1975, a gradual erosion of *Wi Parata* in relation to the existence of native title started to occur.[29]

19 [1877] 3 NZ Jur (NS) 72.
20 Ibid 77.
21 Ibid 78.
22 [1901] AC 561.
23 Ibid 577.
24 *Wallis v Solicitor-General, Protest of Bench and Bar* (1903) 1 NZLRLC 84.
25 *Hohepa Wi Neera v Bishop of Wellington* (1902) 21 NZLR 655 (CA); *Waipapakura v Hempton* (1914) 33 NZLR 1065; *Willoughby v Waihopi* (1910) 13 GLR 41.
26 [1962] NZLR 600.
27 [1963] NZLR 461.
28 Miller et al, above n 12, 227.
29 *Te Weehi v Regional Fisheries Officer* [1986] 1 NZLR 680; *Te Rūnanga o Muriwhenua Inc v Attorney-General* [1990] 2 NZLR 641; *Te Rūnanganui o Te Ika Whenua Inc Society v Attorney-General* [1994] 2 NZLR 20; and *McRitchie v Taranaki Fish and Game Council* [1999] 2 NZLR 139.

Then came the landmark decision of *Attorney-General v Ngati Apa* ('*Ngati Apa*'),[30] in which the Court of Appeal overruled *Ninety-Mile Beach*, finding that native title could potentially exist in the foreshore and seabeds.[31] The New Zealand Parliament's response to *Ngati Apa* was to enact the controversial *Foreshore and Seabed Act 2004* (NZ) ('*FS Act*'),[32] the object of which was to 'preserve the public foreshore and seabed in perpetuity as the common heritage of all New Zealanders'.[33] This object was to be given effect 'by vesting the full and legal beneficial ownership of the public foreshore and seabed in the Crown',[34] thus effectively extinguishing all native title rights in the seabed and foreshore.

However, as the *FS Act* only extinguished native title rights in the foreshore and seabed, it left open the possibility that native title might be recognised in other land and waters where there was no plain legislative intention to extinguish. However, because on-shore Māori title has been progressively converted to freehold by the Native Land Court, it is not immediately apparent where this might be. The *FS Act* has since been repealed.[35]

Although *Ngati Apa* involved the foreshore and seabed, its significance in relation to inland waters lies in statements made in obiter by the judges about the common law recognition of native title in Aotearoa New Zealand.[36] As discussed in Chapter 4, Australian courts have been clear that the common law will only recognise those native title rights and interests that are *capable* of being recognised by the common law and do not 'fracture a skeletal principle of our legal system'.[37] Thus, in Australia, the common law will not recognise exclusive ownership or control of water by native title holders, as the common law does not recognise ownership of water by anyone due to its nature as *res communes* or *publici juris*. The view that water is *publici juris* and therefore incapable of ownership has been accepted in Aotearoa New Zealand in at least one decision.[38] This presumption has been (and continues to be) the position of the Crown. However, the court in *Ngati Apa* cautioned against such presumptions, stating:

> The common law as received in New Zealand was modified by recognised Māori customary property interests. If any such custom is shown to give interests in foreshore and seabed, there is no room for a contrary presumption

30 [2003] 3 NZLR 643.

31 The case did not find that native title existed; however, it held that the legislation did not prevent the continued existence of native title, and therefore the Māori Land Court had jurisdiction to hear a claim.

32 Jacinta Ruru, 'Claiming Title in the Foreshore and Seabed' in Louis A Knafla and Haijo Westra (eds), *Aboriginal Title and Indigenous Peoples: Canada, Australia and New Zealand* (UBC Press, 2010) 190–4; Williams, above n 16, 43–4.

33 *FS Act* s 3.

34 Ibid s 4(a).

35 *Marine and Coastal Area (Takutai Moana) Act 2011* (NZ) s 5.

36 *Ngati Apa* [2003] 3 NZLR 643, 657, 661, 668 (Elias CJ).

37 *Mabo* (1992) 175 CLR 1, 43 (Brennan J).

38 *Glenmark Homestead Limited v North Canterbury Catchment Board* [1975] 2 NZLR 71, 81 (Macarthur J).

> derived from English common law. The common law of New Zealand is different.[39]

Ruru suggests that statements such as this indicate that a New Zealand court may be prepared to consider a departure from the English common law approach in relation to water if a future case was to come before it.[40] In that regard, she notes that given that exclusive ownership is at the extreme end of the spectrum of native title, it is feasible that the New Zealand courts might countenance the recognition of a lesser form of native title in water, for example, of the nature recognised in Australia under the *NTA*. [41]

However, if the New Zealand courts were to look to Australian cases decided under the *NTA* for guidance on how to resolve native title claims to water, they may take a different direction, given the difference in wording of the statutes involved. In the leading Australian case, *Western Australia v Ward*, [42] the 'right to the use and flow and to the control of the water' in s 4(1) of the *Rights in Water and Irrigation Act 1914* (WA) did not extinguish native title rights to water; however, it precluded any exclusivity in water.[43] In the equivalent New Zealand legislation, the *Water and Soil Conservation Act 1967* (NZ) ('*WSC Act*') s 21 provided that

> the sole right to dam any river or stream, or to divert or take natural water, or discharge natural water or waste into any natural water, or to use natural water, is hereby vested in the Crown subject to the provisions of this Act ...[44]

The important difference between the two statutes is the reference in the *WSC Act* to the 'sole right' being vested in the Crown, the effect arguably being that the use of the word 'sole' precludes water rights in anyone but the Crown (subject to the provisions of the Act). It has been suggested that, because of this, an argument for full extinguishment may have greater chance of success in New Zealand.[45] Its success, however, would depend on whether this amounted to a clear and plain intention to extinguish.

Some doubt has been expressed about the likelihood of Māori bringing a case before the courts to test the existence and extent of native title rights in water, due to the cost and time involved and the uncertainty of outcome, given that

39 *Ngati Apa* [2003] 3 NZLR 643, 668 [86]. See also Jacinta Ruru, 'Undefined and Unresolved: Exploring Indigenous Rights in Aotearoa New Zealand's Freshwater Legal Regime' (2009) 20 *Journal of Water Law* 236, 242.

40 Jacinta Ruru, 'The Legal Voice of Māori in Freshwater Governance – A Literature Review' (Landcare Research New Zealand, October 2009) 82.

41 Ruru, 'Undefined and Unresolved', above n 39, 242.

42 (2002) 213 CLR 1.

43 Similar vesting provisions are found in most of the Australian states and territories, except South Australia: see Chapter 4 n 92.

44 The Crown's rights in s 21 were continued in the *RMA* in s 354(1)(b).

45 Meredith Gibbs and April Bennett, 'Māori Claims to Ownership of Freshwater' (2007) *Resource Management Journal* 13, 16.

Parliament may simply legislate to override a favourable court decision (as it did with the *FS Act* following the decision in *Ngati Apa*).[46] This may be unduly pessimistic, as Māori have persistently turned to the courts to recognise their interests in water bodies since the earliest days of English colonial rule.[47] Nonetheless, there is support for this view in the 2012 Waitangi Tribunal's *Freshwater and Geothermal Report*, in which the claimants submitted that 'they are not advancing a native or aboriginal title claim because, in their view, the effect of such a claim at common law is uncertain'.[48] This submission was made in the context of lodging a *Treaty* claim before the Waitangi Tribunal, but it would be equally applicable to the lodging of a claim in the courts.

Concluding comments on native title

Because the issue of native title rights to water in Aotearoa New Zealand is so bound up in the question of ownership (and what is meant by ownership), native title as an avenue for Māori to obtain greater recognition of their rights to participate in the management of water resources is presently quite unsettled and therefore of limited utility for Māori. Native title was also found to be of limited utility for Victorian Traditional Owners, being confined essentially to rights to take and use water for traditional purposes. In both jurisdictions, native title has been and continues to be an inadequate avenue for recognising Indigenous rights to participate in water management.

Given native title's inadequacies, we now turn to an examination of Aotearoa New Zealand's statutory water management regime, contained in the *Resource Management Act 1991* (NZ) ('*RMA*'), to see whether it fares any better.

Māori water rights and the *Resource Management Act*

Introduction

The considerable Māori involvement in the development of the *RMA* and subsequent recognition of Māori interests in its provisions were noted in Chapter 8.

This chapter concentrates on four key aspects of the *RMA* by which it recognises Māori interests and facilitates Māori participation in the management of

46 Ruru, 'Undefined and Unresolved', above n 39, 242.

47 For example, the litigation relating to the Whanganui River has been described as the longest legal battle in New Zealand's legal history: Whanganui Iwi and the Crown, *Ruruku Whakatupua Te Mana o Te Iwi o Whanganui* (5 August 2014) 35. Numerous cases have been (and continue to be) heard under the *RMA*; see Ruru, 'The Legal Voice of Māori', above n 40, 23–48 for a summary of the key cases up to 2009. Other more recent significant Supreme Court cases include *New Zealand Māori Council v Attorney-General* [2013] 3 NZLR 31; *Paki v Attorney-General (No 2)* [2015] 1 NZLR 67.

48 Waitangi Tribunal, *Stage 1 Report on the National Freshwater and Geothermal Resources Claim* (2012) 92.

water resources. Attention is first given to part 2, which contains the overarching principles and purposes of the *RMA* and requirements for *RMA* decision-makers to consider Māori interests. Following is an analysis of provisions that provide for direct participation in decision-making, namely transfers of powers and joint management agreements. Third is an analysis of consultation requirements, and, finally, I undertake an analysis of the use of orders relating to heritage protection and water conservation as an avenue for Māori to participate in water management. The background and context of these provisions will be explained in turn as the chapter proceeds. In summary, each purportedly facilitates Māori participation in water management, but some have not lived up to their initial promise, or have not been of much assistance to Māori to date.

The administrative framework

To contextualise the analysis sketched out above, it is first necessary to outline the basic administrative framework of decision-making under the *RMA* as it relates to water.

The *RMA* controls the use and management of water resources by restricting certain uses of beds of lakes and rivers,[49] by regulating the extraction of water[50] and by regulating discharges into water.[51] Generally speaking, apart from some exceptions,[52] no one can use or affect water resources unless expressly allowed to under a resource consent, national environmental standard, regulation or rule in a regional plan.

The key entities with responsibility for the management of water resources are regional councils.[53] But territorial authorities,[54] the Minister for the Environment, the Minister of Conservation and the Minister for Aquaculture also have a role to play.[55]

The functions of regional councils are set out in s 30 and insofar as they relate to water include: the control and use of land for the maintenance and enhancement of the quality and quantity of water and ecosystems;[56] the control of the taking, use, damming and diversion of water and the quantity, level and flow of water;[57] the control of discharges or contaminants into water;[58] the establishment in regional plans for rules for the taking and use of water;[59] and the control of any planting on the beds of water bodies.[60]

49 *RMA* s 13.
50 Ibid s 14.
51 Ibid s 15.
52 Ibid s 14(3)(b), (c), (e).
53 Ibid s 2 (definition of 'regional council').
54 Ibid s 2 (definition of 'territorial authority').
55 Ibid ss 24–29A.
56 Ibid s 30(1)(c)(ii)–(iiia).
57 Ibid s 30(1)(e).
58 Ibid s 30(1)(f).
59 Ibid s 30(1)(fa).
60 Ibid s 30(1)(g).

The role of territorial authorities in relation to water is limited to 'the control of any actual or potential effects of activities in relation to the surface of water in rivers and lakes'.[61]

For a person to undertake any activity that relates to water or a water body, an application must be made to the relevant local authority,[62] usually the regional council, for a resource consent. Subject to ss 33 and 36B discussed below, Māori are not part of the authority which makes the decision as to whether or not to grant a resource consent. But that does not mean that Māori interests are not relevant to the decision-making process undertaken by local authorities, or that they have no role to play. So let us now turn to our attention to the first aspect under analysis, part 2, and see what it says about the recognition and consideration of Māori interests.

Part 2 of the RMA *– purpose, principles and Māori recognition*

Part 2 consists of only four sections, ss 5–8. The overarching purpose of the *RMA*, contained in s 5(1) is 'to promote the sustainable management of natural and physical resources'. Sustainable management is defined (in part) as:

> managing the use, development, and protection of natural and physical resources in a way, or at a rate, which enables people and communities to provide for their social, economic, and cultural well-being and for their health and safety ...[63]

People and communities includes Māori people and communities,[64] so providing for Māori social, economic and cultural well-being is an inherent part of the overarching purpose of sustainable development. Sections 6 to 8 then set out how this overarching purpose is to be achieved, and it is here that we find specific references to Māori interests.

Under s 6, persons exercising functions or powers under the *RMA* 'shall recognise and provide for' matters of national importance, including

> (e) the relationship of Māori and their culture and traditions with their ancestral lands, water, sites, waahi tapu, and other taonga; (f) the protection of historic heritage[65] from inappropriate subdivision, use, and development; and (g) protection of protected customary rights.

61 Ibid s 31(1)(e).
62 Defined in s 2 of the *RMA* as follows: '**local authority** means a regional council or territorial authority'.
63 Ibid s 5(2).
64 Janet Stephenson, 'Recognising Rangatiratanga in Resource Management for Māori Land: A Need for a New Set of Arrangements?' (2001) 5 *New Zealand Journal of Environmental Law* 159, 186.
65 Defined in s 2 of the *RMA* as including '(b)(iii) sites of significance to Māori, including wāhi tapu'.

Pursuant to s 7, those persons 'shall have particular regard to' other matters, including '(a) kaitiakitanga'. Finally, under s 8 they 'shall take into account' the principles of the *Treaty of Waitangi.*

There is a hierarchy of language between ss 6, 7 and 8 which parallels the order in which they are set out, with 'recognise and provide for' in s 6 being the strongest and 'take into account' in s 8 being the weakest.[66]

However ss 6–8 are all subordinate to the overall purpose of the *RMA* in s 5, which is to promote the sustainable management of natural and physical resources. This was highlighted by Whiting J in the 2005 case of *Genesis Power Ltd v Franklin District Council* ('*Genesis Power*').[67]

Whiting J refers in *Genesis Power* to a 'balancing exercise'.[68] This balancing exercise is required not only between the hierarchy of factors in ss 6–8 but also within s 5 itself.[69] For example, the well-being of Māori communities may need to be balanced against the well-being of other communities; economic well-being may need to be balanced against social and cultural well-being. Additionally, it has been argued that these matters are all subordinate to the matters set out in (2)(a)–(c) which themselves involve a balancing exercise.[70] The case of *Te Rūnanga o Taumarere v Northland Regional Council* illustrates both the fact that Māori people and communities clearly fall within the ambit of s 5 and the balancing required within that section.[71]

These provisions all relate to decision-making by the relevant authorities under the *RMA*. So although they do not guarantee a direct role for Māori in resource management decisions under the *RMA*, they are a means by which Māori interests can influence decision-making and therefore influence the management of water resources.

There are, however, some direct avenues in the *RMA* which theoretically provide the opportunity for Māori participation in decisions relating to the management of water resources, namely ss 33 and 36B.

Transfers of powers – section 33

Local authorities are able to transfer functions, powers and duties to *iwi* authorities by virtue of s 33. When the *RMA* was enacted, this was arguably 'the most significant step in the direction of recognising tino rangitiratanga'.[72]

66 *Takamore Trustees v Kapiti Coast District Council* [2003] 3 NZLR 496, [89] (Ronald Young J). See also *Haddon v Auckland Regional Council* [1994] NZRMA 49, 60–61.
67 [2005] NZRMA 541.
68 Ibid [55].
69 For a detailed analysis of s 5, see Nicola Wheen, 'The Resource Management Act 1991: A Greener Law for Water?' (1997) 1 *New Zealand Journal of Environmental Law* 165, 176–98.
70 Ibid 183–94.
71 [1996] NZRMA 77, 91–2.
72 Stephenson, above n 64, 187.

However, despite its promise, this section has been of little utility in enhancing Māori decision-making over water resources. To date there have been no transfers of any functions, powers or duties to *iwi* authorities under s 33.

Various barriers have been identified as inhibiting transfers under this section,[73] one being that instead of using s 33, 'councils have preferred to address iwi concerns using other more familiar, less powerful mechanisms, such as contracting and delegation (both leaving the powers largely in the hands of councils)'.[74] These alternative mechanisms are useful for building better relationships, but as it was further argued:

> without these legislative requirements [ie ss 6–8] to involve tāngata whenua in resource management, most local authorities would still not have established any form of relationship with them. It is legislative requirement that has brought the relationship to this point.[75]

This highlights the importance of legislative recognition, as a facilitator at least, for establishing relationships and therefore improving the likelihood of better engagement with Indigenous people in decision-making about water resources.

Another barrier identified was the difficulty in defining an *iwi* authority. In that regard, the Whanganui River Settlement discussed in the next chapter specifically designates the post-settlement governance entity ('PSGE') as an *iwi* authority for the purposes of the *RMA*, [76] so there would be no question about the status of the PSGE should it decide to pursue a transfer of powers under s 33.

The Treaty claims settlement process and the consequential establishment of PSGEs, usually with some settlement resources at their disposal for operational and capacity-building purposes, may therefore awaken s 33 from its slumber. The existence of registered native title bodies corporate ('RNTBCs') and traditional owner group entities ('TOGEs') in Victoria, with similar settlement resources at their disposal, suggests that this concept, although currently unutilised in Aotearoa New Zealand, may be worth considering by Victorian Traditional Owners. Further afield, an ability to transfer powers to Indigenous entities already exists in the USA which enables federally recognised tribes to be 'treated as a state'[77] by the Environment

73 Hamish Rennie, Jill Thomson and Tikitu Tutua-Nathan, *Factors Facilitating and Inhibiting Section 33 Transfers to Iwi* (Geography Department, University of Waikato, 2000). See also Elizabeth Clark, 'Section 33 of the *Resource Management Act 1991*' in Janine Hayward, *Local Government and the Treaty of Waitangi* (Oxford University Press, 2003) 49–50.

74 Rennie, Thomson and Tutua-Nathan, above n 73, 3.

75 Ibid 53–4.

76 *Te Awa Tupua (Whanganui River Claims Settlement) Act 2017* (NZ) s 72.

77 *Safe Drinking Water Act* 42 USC § 300j-11 (1986) ('*SDWA*'); *Clean Water Act* 33 USC § 1377(e) (1987). See Environment Protection Agency, 'Tribal Assumption of Federal Laws – Treatment as a State (TAS)' <www.epa.gov/tribal/tribal-assumption-federal-laws-treatment-state-tas>.

Protection Agency for various environmental programmes (although its uptake has been mixed).[78]

Joint management agreements – section 36B

Another opportunity for Māori participation in decision-making about water resource management is via joint management agreements ('JMAs') under s 36B, an amendment made to the *RMA* in 2005 intended to encourage and build on previous successful partnerships between councils and Māori.[79]

Under s 36B, an *iwi* authority, a group which represents *hapū* interests, or a public authority, can enter into a JMA with a local authority provided certain conditions are satisfied.[80] Those conditions include that it 'represents the relevant community of interest',[81] 'has the technical or special capability or expertise'[82] to perform the relevant functions, and that the JMA 'is an efficient method of performing' the relevant functions.[83] The JMA must also include details of the resources required for the administration of the JMA and how those administrative costs will be met.[84]

It was noted in 2007 that 'the reluctance by Councils to transfer powers to date [ie under s 33] does not bode so well for these proposed joint management agreements'.[85] This prediction appears to have been borne out in practice.

The first JMA made under s 36B was between the Tūwharetoa Māori Trust Board and the Taupō District Council in January 2008.[86] But it appears that this is also the only JMA to have been reached pursuant to s 36B.[87]

78 The Navajo Nation is the only tribe to date to have been granted TAS and primacy under the *SDWA* for the Public Water System Supervision program: <www.epa.gov/tribaldrinkingwater/tribal-governments-role-safe-drinking-water-tribal-lands#tab-2>. On the other hand, under the *Clean Water Act*, 58 tribes, including the Navajo Nation, have been granted TAS status to administer water quality standards (WQS) programmes on their lands, of which 44 have had their WQS approved by the EPA: <www.epa.gov/wqs-tech/epa-approvals-tribal-water-quality-standards-and-contacts>. For more detail on the Navajo Nation, see Jill Elise Grant, 'The Navajo Nation EPA's Experience with "Treatment as a State" and Primacy' (2007) 21(3) *Natural Resources & Environment* 9.

79 New Zealand, *Parliamentary Debates*, 2 August 2005, 22273 (David Benson-Pope, Associate Minister for the Environment).

80 *RMA* s 36B(2).

81 Ibid s 36B(1)(b)(i)(A).

82 Ibid s 36B(1)(b)(i)(B).

83 Ibid s 36B(1)(b)(ii)

84 Ibid s 36B(1)(c)

85 Prue Kapua, 'Review of the Role of Māori Under the *Resource Management Act 1991*' (2007) *Resource Management Theory and Practice* 92, 107.

86 Taupō District Council, *Joint Management Agreements* (4 June 2014) <www.taupodc.govt.nz/our-council/policies-plans-and-bylaws/joint-management-agreements/Pages/joint-management-agreements.aspx>.

87 Samuel George Wevers, 'Recognising Rangatiratanga Through Co-management: The Waikato River Settlement' [2013] 4 *New Zealand Law Review* 689, 711.

So although s 36B was clearly a step forward in providing the potential for management opportunities by Māori over water resources, its lack of uptake suggests that it has not proven to be a particularly viable avenue in practice.

The existence of s 36C may also potentially inhibit the establishment of s 36B JMAs. This section provides that, in certain circumstances, a local authority can perform a function, power or duty that is to be performed jointly under the JMA, without the involvement of the other joint management party.[88] This gives a local authority the ability to make unilateral decisions under a JMA, which is contrary to the underlying purpose of a JMA, that being joint decision-making. This diminishes the incentive for an *iwi* authority or group representing *hapū* interests to enter into a JMA. On the other hand, it might reduce the concerns of local authorities because it means that they retain some ability to make urgent decisions without the need to involve the joint management partner.

A JMA can also be unilaterally terminated under s 36E. This is a further disincentive for Māori to enter into a JMA as termination by either party will simply result in decision-making reverting back to the local authority.[89]

Rather than use JMAs under s 36B, it has been more effective for Māori to negotiate joint management arrangements through other mechanisms such as the *Treaty* settlement process, a prime example being the co-management arrangements in the Waikato River Settlement, discussed in next chapter.[90] The main advantage of using the *Treaty* settlement process is that the co-management arrangements are usually set out in the settlement legislation (that is, they are mandatory), thus 'eliminating the problem of reluctance by councils to enter into a dual management type of agreement with iwi'.[91]

What the above discussion illustrates is that the two provisions of the *RMA* which allow for direct participation in resource management (and therefore water management), namely ss 33 and 36B have not lived up to their initial promise.

Consultation and resource consents – section 36A

Along with s 36B, included in the amendments made to the *RMA* in 2005 was the insertion of s 36A,[92] which provides that there is no duty on local authorities or applicants to consult about resource consent applications. This was essentially a confirmation of 'the existing legal position that neither an applicant nor a consent

88 These circumstances are limited to where time is of the essence and there is no method in the JMA for the making of decisions of that kind: *RMA* s 36C(2).

89 Natalie Coates, 'Joint Management in New Zealand: Simply Empty Promises?' (2009) 13(1) *Journal of South Pacific Law* 32, 34.

90 See also Local Government New Zealand, *Co-management: Case Studies Involving Local Authorities and Māori* (January 2007) and Local Government New Zealand, *Local Authorities and Māori: Case Studies of Local Arrangements* (February 2011).

91 Coates, above n 89, 37.

92 For a detailed discussion of s 36A, see Jenny Vince, 'Māori Consultation Under the *Resource Management Act* and the 2005 Amendments' (2006) 10 *New Zealand Journal of Environmental Law* 295.

authority has a duty to consult any person [including Māori] in respect of applications for resource consents and notices of requirement'.[93]

So while the inclusion of s 36B (in theory) improved the potential for Māori participation in water management, s 36A was clearly designed to preserve the status quo, whereby Māori are excluded from participating in decision-making.

Consultation may still occur,[94] and must still occur if required by another Act.[95] It has been suggested[96] that this latter requirement is important because it preserves consultation requirements under the *Local Government Act 2002* (NZ) ('*LGA*').[97]

In relation to the involvement of Māori in the resource consent approval process, the only formal way that Māori can have some direct influence on water management in the absence of consultation is by way of a third-party challenge to a resource consent. This, however, comes with its own problems, not least being the time and resources required to monitor the granting of consents, assess the impacts and pursue a court challenge.[98] And in that regard, the Māori experience in court has been one of only limited success in relation to water.[99]

Consultation and policy statements and plans

Section 36A relates specifically to resource consent applications. On the other hand, requirements for the Minister to consult with Māori in relation to the preparation of national policy statements and for local authorities to consult with Māori in relation to the preparation of policy statements and plans were improved by the 2005 amendments to the *RMA*. [100] To assist local authorities to know with whom to consult, amendments were also made whereby the details of the relevant *iwi* authorities are entered into a record kept by the local authority.[101] This would also assist in any prudent (but not compulsory) consultation undertaken by the local authority or applicant for a resource consent application. An amendment to the *RMA* in 2017 now requires local authorities, prior to any public notification, to provide a copy of any draft policy statement or plan to any *iwi* authorities

93 Linda Te Aho, 'Contemporary Issues in Māori Law and Society' (2005) 13 *Waikato Law Review* 145, 163. The existing legal position is reflected in *Carter Holt Harvey v Te Rūnanga o Tūwharetoa Ki Kawerau* [2003] 2 NZLR 349, [55] (Heath J).

94 *RMA* s 36A(1)(c).

95 Ibid s 36A(1)(b).

96 Vince, above n 92, 302, n 29; Derek Nolan (ed), *Environmental and Resource Management Law* (LexisNexis, 5th edn, 2015) [14.32].

97 See, eg, *LGA*, pt 6 and particularly ss 81 and 82(2).

98 James Douglas Kahotea Morris, *Affording New Zealand Rivers Legal Personality: A New Vehicle for Achieving Māori Aspirations in Co-management?* (LLM Thesis, University of Otago, 2009) 69–70.

99 Ruru, 'Undefined and Unresolved', above n 39, 238–40; Ruru, 'The Legal Voice of Māori', above n 40, 23–49.

100 See, eg, s 46, sch 1 cl 3B.

101 *RMA* s 35A.

previously consulted, provide adequate time and opportunity to comment, and have particular regard to any advice received.[102]

The *RMA* was also amended in 2009 to require Māori participation in the preparation of national environmental standards.[103] The Minister must now notify *iwi* authorities of the proposed subject matter of the standard, and establish a process which gives *iwi* authorities adequate time and opportunity to comment on the proposed subject matter of the standard.[104] Although the obligation to provide an adequate opportunity to comment is weaker than an obligation to consult,[105] this is nonetheless a further, albeit limited, addition to the various mechanisms in the *RMA* for Māori input into water management, insofar as a national environmental standard relates to water.

Because there is no requirement for either the applicant for a resource consent or the local authority to which the application is made to consult with *tāngata whenua* [106] in relation to the application, it is important that Māori make the most of the consultations that are required in relation to any standards, policy statements and plans. This is particularly so given that it is a requirement of the *RMA* that consent authorities have regard to these documents when considering resource consent applications.[107] As one commentator has noted, '[i]n most cases, reference to these documents is the first stage for resource consent applicants in determining if their development would affect sites or Māori interests'.[108]

By way of example, the *National Policy Statement for Freshwater Management* ('*NPS-FM*') discussed in Chapter 8 has specific sections on *tāngata whenua* roles and interests,[109] as well as numerous other references to Māori interests.[110] Both regional plans and district plans must give effect to any national policy statements,[111] including the *NPS-FM*. So a consent authority must have regard not only to the *NPS-FM*, but also to the regional and district plans which are required to give effect to the *NPS-FM*. Given the requirements of the *NPS-FM* in relation to the recognition of Māori interests in water, a consent authority cannot avoid having regard to those interests.

In a 2008 report on Māori provisions in council planning documents, it was concluded 'that *RMA* policies and plans can be characterised as having fairly

102 Ibid sch 1 cl 4A.
103 *Resource Management (Simplifying and Streamlining) Amendment Act 2009* (NZ).
104 *RMA* s 44(2).
105 *New Zealand Fishing Industry Association Inc v Minister of Agriculture and Fisheries* [1988] 1 NZLR 544, 551. This case referred to a right to make submissions (under the *Fisheries Act 1983* (NZ)), which is essentially the same as an opportunity to comment. See also *Harris v Great Barrier Reef Marine Park Authority* (2000) 98 FCR 60, 67.
106 Defined in the *RMA* s 2 as follows: '**tāngata whenua**, in relation to a particular area, means the iwi, or hapū, that holds mana whenua over that area'.
107 *RMA* s 104(1)(b).
108 Vince, above n 92, 297.
109 *NPS-FM 2014* ss AA, D.
110 Ibid 4–5, 7, appendix 1.
111 *RMA* ss 67(3)(a), 75(3).

strong Mäori provisions'.[112] However, this conclusion was qualified, the author noting that

> these [provisions] largely paraphrase the Mäori provisions of the *RMA* itself and consequently often fails to address local circumstances for Mäori. Another widely observed deficiency is that plans include strong high-level recognition of Mäori values and issues, but fail to translate these into effective and viable methods that councils and tāngata whenua can use.[113]

There are, however, additional opportunities in the *RMA* for Māori to participate in the management of water resources via the heritage protection and water conservation order provisions. The advantages and disadvantages of these mechanisms are discussed below.

Heritage protection authorities and heritage orders

An *iwi* authority, if a body corporate,[114] can apply to be a heritage protection authority ('HPA') in order to protect any place, which includes 'any feature or area'[115] except for 'any place or area that is private land'.[116] This would encompass water bodies not on private land.

The advantage for an *iwi* authority of being an HPA is that an HPA can require a territorial authority to consider including a heritage order in its district plan (subject to various public notification requirements), one of the purposes of a heritage order being to protect 'any place ... of special significance to the tāngata whenua for spiritual, cultural, or historical reasons',[117] along with any surrounding land if reasonably necessary.[118]

The consequence of including a heritage order in a district plan is that the consent of the *iwi* authority (as the HPA) is then required to undertake any activity that would wholly or partly nullify the effect of the heritage order.[119] Accordingly, if a territorial authority proceeds to include a heritage order relating to a water body in a district plan, this enables the *iwi* authority to have a specific management function of a protective nature, in relation to that water body which is the subject of the heritage order and for which it has HPA status; it effectively becomes a de facto consent authority.

112 Nathan Kennedy, 'Viewing the World Through a Wider Lens: Mäori and Council Planning Documents' (PUCM Mäori Report 6, International Global Change Institute, University of Waikato; and KCSM Consultancy Solutions, 31 July 2008) 18 <http://researchcommons.waikato.ac.nz/handle/10289/6106>.

113 Ibid.

114 *RMA* s 188(1).

115 Ibid s 188(2).

116 Ibid s 188(1A).

117 Ibid s 189(1)(a).

118 Ibid s 189(1)(b).

119 Ibid s 193.

Local authorities, Ministers of the Crown and Heritage New Zealand Pouhere Taonga are automatically HPAs;[120] however, only five entities have ever been approved as an HPA under this section (two of which were subsequently revoked), none of which is a Māori entity.[121]

It has been suggested[122] that the Māori experience with the heritage protection provisions of the *RMA* has been informed by *Te Rūnanga o Ngati Pikiao v Minister for the Environment*, [123] whereby despite the High Court upholding the *iwi*'s judicial review application and ordering the Minister to reconsider his decision, the *iwi* body corporate was still not approved as an HPA.

Before approving an application to become an HPA, the Minister must be satisfied that the applicant is appropriate for the role, and that it can satisfactorily carry out all the responsibilities, including financial responsibilities, of being an HPA.[124]

Some Māori organisations may find it difficult to satisfy the Minister that they can carry out the financial responsibilities required. Further, the knowledge that being approved as an HPA carries with it what could potentially be significant financial responsibilities may act as a disincentive to putting in an application in the first place.

So, although this is a further avenue by which Māori can participate in the management of water resources, as with 33 and 36B it has proven to be of little utility. However, with the increasing number of *Treaty* settlements involving the provision of resources for capacity building and operational costs, it is possible that this may change. In that respect, the Whanganui River Settlement discussed in the next chapter is of particular interest because it specifically contemplates one of the settlement entities applying to become an HPA.[125]

Water Conservation Orders

Water Conservation Orders ('WCOs') made under part 9 of the *RMA* may potentially be another avenue for Māori to participate in water management.

WCOs first made an appearance in 1981 by way of an amendment to the *WSC Act* [126] and then reappeared in the *RMA*. Importantly, the *RMA* now contains, in s 199, the purpose of a WCO, which is

> to recognise and sustain—

120 Ibid s 187(a), (b), (c).

121 See Ministry for the Environment, 'Heritage Protection Authorities' (22 August 2017) <www.mfe.govt.nz/rma/about-rma/roles-and-responsibilities/heritage-protection-authorities>.

122 Nolan, above n 96, [14.34]; Paul F Majurey et al, *Māori Values Supplement* (Ministry for the Environment, December 2010) 306.

123 (Unreported, High Court of New Zealand, Gallen J, 15 June 1999).

124 *RMA* s 188(5).

125 *Te AwaTupua (Whanganui River Claims Settlement) Act 2017* (NZ) s 17(f).

126 *Water and Soil Conservation Amendment Act 1981* (NZ).

(a) outstanding amenity or intrinsic values which are afforded by waters in their natural state:
(b) where waters are no longer in their natural state, the amenity or intrinsic values of those waters which in themselves warrant protection because they are considered outstanding.

There was no equivalent of s 199 in the *WSC Act.*

A list of outstanding characteristics to be protected by a WCO is included in s 199. Most notably, a WCO may provide for 'the protection of characteristics which any water body has or contributes to, and which are considered to be of outstanding significance in accordance with tikanga Māori'.[127]

Any person can apply to the Minister for a WCO upon the payment of the prescribed fee.[128] However, the making of a WCO can be a complex and lengthy process.[129] For example, the application for the Buller WCO was filed in September 1987 with gazettal finally occurring in June 2001, nearly 14 years later.[130] One of the shortest time frames for making a WCO was in relation to the Oreti River WCO, which still took approximately three and a quarter years.

There are currently 15 WCOs in force (along with six amendment WCOs), seven of which were made pre-*RMA* (ie under the *WSC Act*).

None of the WCOs under the *WSC Act* (as originally gazetted)[131] protect waters with characteristics of outstanding significance to Māori, reflecting the lack of detailed purposes and the more limited scope of matters to be taken into account under the *WSC Act.* [132] Only four WCOs made under the *RMA* protect waters which have outstanding characteristics of significance for tikanga Māori. These are the two most recent WCOs relating to the Rangitata and Oreti Rivers,[133] the third being the Kawarau WCO and the fourth was by way of an amendment to the Te Waihora/Lake Ellesmere WCO.[134] The Te Waihora/Lake Ellesmere amendment is particularly notable because it involved a Māori applicant.[135]

127 *RMA* s 199(2)(c). Tikanga Māori is defined in s 2 of the *RMA* as 'Māori customary values and practices'.
128 Ibid s 201(1).
129 *RMA* ss 201–15. See, eg, *Talley v Fowler* (Unreported, High Court of New Zealand, Fogarty J, 18 July 2005) [24].
130 *Talley v Fowler* (Unreported, High Court of New Zealand, Fogarty J, 18 July 2005) [25].
131 The Te Waihora/Lake Ellesmere WCO was amended in 2011 to include outstanding values of 'significance in accordance with tikanga Māori in respect of Ngāi Tahu history, mahinga kai and customary fisheries'.
132 *WSC Act* ss 20B(6), 20F(7).
133 The *Water Conservation (Rangitata River) Order 2006* (NZ) and the *Water Conservation (Oreti River) Order 2008* (NZ).
134 *National Water Conservation (Te Waihora/Lake Ellesmere) Order 2011* (NZ).
135 This was a joint application by Te Rūnanga o Ngāi Tahu and the Department of Conservation.

However, none of the other applications for WCOs, either under the *WSC Act* or the *RMA*, were lodged by Māori. Almost two-thirds of the applications were made by various Fish and Game Councils (previously Acclimatisation Societies).[136]

The lengthy and complex WCO process would be a significant disincentive to making an application, but this does not mean a lack of involvement by Māori; it is clear from the written record that Māori have participated in the submission/hearing process for at least four WCO applications.[137] But as WCO applications have generally been made by Fish and Game Councils, it is only those rivers which are valued primarily for recreational fishing that obtain protection by a WCO.[138] Accordingly, the protection of Māori values is only a secondary consequence.

In addition to the problems with the WCO process, the threshold for a successful application is very high; the purpose of WCOs being 'to recognise and sustain … outstanding amenity or intrinsic values'.[139] In *Rangitata South Irrigation Limited v New Zealand Fish and Game* ('*Rangitata South*'), the Environment Court provided some guidance as to the meaning of 'outstanding', which supports the high threshold for a successful application.[140]

The application fee is also relatively high, around $1,022,[141] but this pales in comparison with the cost of pursuing an application. For example, the application for the Rangitata River WCO cost the applicants $543,000 in legal fees and expert advice.[142] When combined with the high threshold for a WCO, these kinds of costs would be a significant deterrent to making an application.

It is therefore not surprising that only 15 WCOs and six amendments have been made since they were first introduced more than 30 years ago, and that only one, that one being an amendment, involved a Māori applicant.[143]

136 Thirteen out of 21 applications (including amendment applications).

137 *Re National Water Conservation (Mohaka River) Order 1990* (Unreported, Planning Tribunal, Sheppard J, 8 April 1992); *Rangitata South* (Unreported, Environment Court of New Zealand, Jackson J, Environment Commissioner Manning and Deputy Environment Commissioner Grigg, 5 August 2004); *Report by a Special Tribunal Appointed by the Minister for the Environment to Consider an Application for a Water Conservation Order for the Oreti River* (November 2007); *Application for a Water Conservation Order on Lake Ellesmere – Report of the Committee of the National Water and Soil Conservation Authority* (1988): Ministry for the Environment, 'Existing Water Conservation Orders' <www.mfe.govt.nz/fresh-water/reform-programme/water-conservation-orders/existing-water-conservation-orders>.

138 Parliamentary Commissioner for the Environment, *Hydroelectricity or Wild Rivers? Climate Change Versus Natural Heritage* (May 2012) 32.

139 *RMA* s 199(1)(a).

140 (Unreported, Environment Court of New Zealand, Jackson J, Environment Commissioner Manning and Deputy Environment Commissioner Grigg, 5 August 2004) [17].

141 Contrast this with the fee of around $255 for an application to become an HPA: *Resource Management (Forms, Fees and Procedure) Regulations 2003* (NZ) sch 2.

142 *Hydroelectricity or Wild Rivers?*, above n 138, 32.

143 However, as at February 2018 there were three WCO applications, two of which involve a Māori applicant: Ministry for the Environment, 'Current Water Conservation Order Applications' (15 September 2017) <www.mfe.govt.nz/freshwater/water-conservation-orders/current-applications>.

But there are other disincentives for Māori to use the WCO process as a way of participating in the management of water resources. Importantly, a WCO does not give any specific management rights to Māori. What it means is that a consent authority cannot issue a resource consent that is contrary to any provision of the WCO, or, if granting a resource consent, must ensure that appropriate conditions are in place to maintain the provisions of the WCO.[144]

Second, the imposition of a WCO does not affect any pre-existing resource consents.[145] So if there are existing resource consents relating to the water body that are affecting outstanding characteristics of significance in accordance with tikanga Māori, then applying for a WCO will not assist.

Nonetheless, if Māori are not deterred by the time and expense of making an application for a WCO, a WCO may still be useful in that it may not then be necessary for Māori to object to future applications for resource consents in relation to the relevant water body for it to be protected. WCOs are also one of the strongest measures in the *RMA* for protecting water bodies, because they take precedence over regional policy statements, regional plans and district plans.[146] In addition, a WCO that is more stringent than a national policy statement ('NPS') will prevail over the NPS.[147]

Finally, the WCO provisions also take precedence over the provisions of part 2. This emphasises the status of WCOs as being at the top of the resource management hierarchy in the *RMA*, but it also reflects the primacy given to the conservation purpose of WCOs,[148] a point which was made in *Rangitata South.* [149] So although Māori characteristics of a water body can be protected by a WCO, it is for the purpose of conservation rather than sustainable development in part 2. That is not to say that part 2 is irrelevant or can be ignored, but where there are competing provisions, the WCO provisions take precedence.[150]

WCOs are clearly a very strong form of protection for water bodies, but they have significant limitations for Māori in that they do not provide for direct participation in the management of the relevant water body; they are focussed on conservation values (which do not necessarily equate to Māori values) and they are expensive, time-consuming and difficult to apply for.

Concluding remarks on the *RMA*

There are various mechanisms in the *RMA* which provide for Māori participation in water resource management and for Māori interests to be considered in various

144 *RMA* s 217(2).

145 Ibid s 217(1). See also *Rangitata South* (Unreported, Environment Court of New Zealand, Jackson J, Environment Commissioner Manning and Deputy Environment Commissioner Grigg, 5 August 2004) [49]–[50].

146 See *RMA* ss 62(3), 67(4) and 75(4).

147 Ibid s 43C.

148 Nolan, above n 96, [8.70].

149 (Unreported, Environment Court of New Zealand, Jackson J, Environment Commissioner Manning and Deputy Environment Commissioner Grigg, 5 August 2004) [23].

150 Ibid [21]–[25]; *Re National Water Conservation (Karawau) Order 1993* (Planning Tribunal, Skelton J, 13 June 1996) 20; Nolan, above n 96, [8.70].

decision-making processes. However, experience to date suggests that the practical impact of these mechanisms, particularly those allowing for direct participation, has been minimal. This should not be surprising given that they are not mandatory. The most effective provisions would appear to be those that are mandatory, namely consultation with *iwi* in the preparation of plans and policy statements. But even these provisions have their limitations, allowing input only at the higher level, rather than in decision-making on the ground. This is similar to the situation under Victoria's water management legislation.

10 *Treaty* settlements

Due to the inadequacies of the *RMA*, Māori have turned to *Treaty* settlements to achieve their aspirations for water management.[1] There are two settlements on which this chapter will focus, namely the Waikato River Settlement and the Whanganui River Settlement. As background to these two settlements, and to indicate how *Treaty* settlements in relation to inland waters have developed over time, it will begin with a brief outline of both the 1998 Ngāi Tahu Settlement as it relates to Te Waihora (Lake Ellesmere) and the 2006 Te Arawa Lakes Settlement.

The Ngāi Tahu Settlement – Te Waihora

The Ngāi Tahu Settlement was one of the earliest *Treaty* settlements involving a water body, the settlement resulting in the *Ngāi Tahu Claims Settlement Act 1998* (NZ) ('*NTCS Act*').[2]

Under that part of the *NTCS Act* which relates to Te Waihora, title to the bed of Te Waihora is vested in Te Rūnanga o Ngāi Tahu.[3] However, it does not include any rights of ownership, management or control of the waters of Te Waihora.[4] Existing public access and use are preserved,[5] as are existing lawful commercial uses and structures.[6]

Importantly, in relation to the management of water resources, Te Rūnanga o Ngāi Tahu can enter into an agreement with the Minister to prepare a joint

1 Linda Te Aho, 'Indigenous Aspirations and Ecological Integrity: Restoring and Protecting the Health and Wellbeing of an Ancestral River for Future Generations in Aotearoa New Zealand' in Laura Westra, Klaus Bosselman and Colin Suskolne (eds), *Globalisation and Ecological Integrity in Science and International Law* (Cambridge Scholars, 2011) 346, 352; Samuel George Wevers, 'Recognising Rangatiratanga Through Co-Management: The Waikato River Settlement' (2013) 4 *New Zealand Law Review* 689, 710–11.

2 The settlement also involved similar arrangements for Muriwai (Coopers Lagoon) and Lake Mahinapua, but without the option of a JMP.

3 *NTCS Act* s 168.

4 Ibid s 171.

5 Ibid s 173.

6 Ibid s 174.

management plan ('JMP') for the integrated management of the bed of Te Waihora and the natural and historic resources of the area.[7] A JMP was approved for Te Waihora in late 2005,[8] which was 'the first statutory joint management plan between the crown and Iwi'.[9]

Compared with more recent settlements, the Ngāi Tahu settlement in relation to Te Waihora is modest as far as management of water resources is concerned, the JMP being the one noteworthy element.

The Te Arawa Lakes Settlement

The Te Arawa Lakes claim was lodged in the Waitangi Tribunal in 1987. Rather than proceed to a Tribunal hearing, Te Arawa negotiated its settlement directly with the Crown.[10] After lengthy negotiations, legislation implementing the settlement was enacted in 2006, namely the *Te Arawa Lakes Settlement Act 2006* (NZ) ('*TALS Act*').

Under the settlement, title to 13 lake beds has been vested in fee simple in the Te Arawa Lakes Trust ('TALT') and held in trust for Te Arawa.[11] Title to the lake beds is inalienable and cannot be mortgaged, although the TALT can grant leaseholds for up to 35 years, and licences and easements for any length of time.[12]

The vesting of the lake beds in the TALT does not create any rights in relation to the water.[13] In that respect, it is like the *NTCS Act* in relation to Te Waihora. This is a recurring theme of all the settlements involving water bodies, that theme being arguably less about the physical distinction between land and water and more about the government's desire to maintain control over water in the public interest through its continuing assertion that there is no ownership of water.

Various activities can be undertaken and existing structures maintained and used without the consent of or charge imposed by the TALT.[14] Again, this is quite similar to the *NTCS Act*.

7 Ibid s 177.
8 Te Rūnanga o Ngāi Tahu and Department of Conservation, *Te Waihora Joint Management Plan – Mahere Tukutahi o Te Waihora* (10 December 2005).
9 Department of Conservation (NZ), *Conservation Management Plans* <www.doc.govt.nz/about-doc/policies-and-plans/conservation-management-plans/te-waihora-joint-management-plan>.
10 Te Arawa and Arawa Māori Trust Board and Her Majesty the Queen in right of New Zealand, *Deed of Settlement of the Te Arawa Lakes Historical Treaty Claims and Remaining Annuity Issues* (18 December 2004) 4.
11 *TALS Act* s 23.
12 Ibid s 24.
13 Ibid s 25.
14 Ibid ss 32 (recreational activities), 33 (existing structures), 36 (existing commercial activities).

New commercial activities (where consent of the owner is required by other legislation) and new structures require the consent of the TALT.[15] The grant of consent can impose conditions, which can include a charge.[16] Generally speaking, public utilities also require the consent of the TALT to construct.[17] Existing public utilities are covered by the section relating to existing structures.

The *TALS Act* also establishes a co-governance entity, the Rotorua Lakes Strategy Group ('RLSG').[18] The purpose of the RLSG is

> to contribute to the promotion of the sustainable management of the Rotorua lakes and their catchments, for the use and enjoyment of present and future generations, while recognising and providing for the traditional relationship of Te Arawa with their ancestral lakes.[19]

It is a permanent group, and cannot be discharged except with the agreement of the organisations represented on it.[20] Two of the six members of the RLSG are appointed by the TALT.[21]

The RLSG has various functions, including the preparation, approval, monitoring and evaluation of agreements, policies and strategies relating to the Lakes, and involvement in various matters relating to significant issues.[22] Although the RLSG has prepared a strategy for the Lakes,[23] neither the strategy nor any of its other functions gives the RLSG a major role in decision-making.

A notable point of difference between the Te Arawa Settlement and the Ngāi Tahu Settlement over Te Waihora is the absence of any provision for the preparation of a JMP for the Lakes.

The settlement has been aptly described as 'rather pedestrian because it does not give any real decision-making power to Te Arawa'.[24] This can be contrasted with our next settlement under review.

15 Ibid s 41.
16 Ibid s 41(3).
17 Although such consent must not be unreasonably withheld, nor can a charge be imposed as a condition of consent: ibid s 39.
18 Ibid ss 47–51.
19 Ibid s 49.
20 Ibid s 48(3)
21 Composition of the RLSG is not set out in the *TALS Act*, but is incorporated by virtue of s 50, which incorporates the Rotorua Lakes Strategy Group Agreement, defined in s 11. It is this agreement which sets out the membership of the RLSG: Te Arawa and Arawa Māori Trust Board and Her Majesty the Queen in right of New Zealand, *Schedules to the Deed of Settlement of the Te Arawa Lakes Historical Claims and Remaining Annuity Issues* (18 December 2004) Relationships Schedule, Part 1, Rotorua Lakes Strategy Group Agreement, cl 2 ('*Relationships Schedule*').
22 *Relationships Schedule*, above n 21, cl 4.
23 Rotorua Te Arawa Lakes Programme, *Vision and Strategy for the Lakes of the Rotorua District* (2013) <www.rotorualakes.co.nz/strategy>.
24 Jacinta Ruru, 'Indigenous Restitution in Settling Water Claims: The Developing Cultural and Commercial Redress Opportunities in Aotearoa, New Zealand' (2013) 22 *Pacific Rim Law and Policy Journal* 311, 334.

The Waikato River Settlement

Background

The Waikato River Settlement was reached by way of direct negotiations, commencing in 1999, with a Deed of Settlement eventually signed in 2008.[25] Following a change in government, the Deed was amended in 2009 to provide 'more effective and economically efficient arrangements for delivering the overarching purpose of the settlement' and in particular to 'streamline the co-governance arrangement'.[26] The key aspects of the amended Deed were:

- a vision and strategy document which will have special and unique legislative status as the primary direction-setting document for the river;
- a single co-governance entity; and
- joint management agreements.[27]

Implementation of the 2009 Deed of Settlement finally occurred with the passage of the *Waikato-Tainui Raupatu Claims (Waikato River) Settlement Act 2010* (NZ) ('*Waikato River Settlement Act*').[28]

The settlement was seen as a milestone in the way it dealt with Māori interests in water management. 'Heralded as revolutionary',[29] it has also been described as a 'bold vision' containing 'groundbreaking provisions', and an 'innovative approach towards managing the country's longest and most economically significant river'.[30]

The Waikato River Settlement built considerably on previous settlements involving water bodies, such as the Ngāi Tahu Settlement regarding Te Waihora and the Te Arawa Lakes Settlement. The three key aspects of the 2009 Deed of Settlement as reflected in the *Waikato River Settlement Act* will be the main focus of our analysis in the next section.

Details of the settlement

The overarching purpose of the Waikato River Settlement is 'to restore and protect the health and wellbeing of the Waikato River for future

25 *Waikato-Tainui Raupatu Claims (Waikato River) Settlement Act 2010* (NZ) preamble (15).
26 Christopher Finlayson, 'Waikato River Deed of Settlement Signed with Waikato-Tainui' (Media Release, 17 December 2009).
27 Ibid.
28 Similar legislation was also enacted for several related settlements: *Ngati Tūwharetoa, Raukawa, and Te Arawa River Iwi Waikato River Act 2010* (NZ).
29 Ruru, 'Indigenous Restitution', above n 24, 334.
30 Jeremy Baker, 'The Waikato-Tainui Settlement Act: A New High Water Mark for Natural Resources Co-management' (2013) 24 *Colorado Journal of International Environmental Law and Policy* 163, 163. See also Linda Te Aho, 'Indigenous Challenges to Enhance Freshwater Governance and Management in Aotearoa New Zealand – The Waikato River Settlement' (2009) 20 *Journal of Water Law* 285.

generations'.[31] This overarching purpose is one of the guiding principles to be used in interpreting the Act, along with the Vision and Strategy; and the agreements expressed in the 2009 Deed and the Kiingitanga Accord.[32]

Vision and Strategy, 'Te Ture Whaimana'

The Vision and Strategy is set out in schedule 2 of the *Waikato River Settlement Act*, giving it legislative status. It 'is intended by Parliament to be the primary direction-setting document for the Waikato River and activities within its catchment affecting the Waikato River'.[33]

The *Waikato River Settlement Act* cements the paramount status of the Vision and Strategy as the key direction-setting document by providing that the Vision and Strategy will prevail over any inconsistent policy or planning document in the *RMA* (including national policy statements),[34] that *RMA* regional policy and planning documents will need to be updated to conform with the Vision and Strategy,[35] and that they contain statements indicating how they have given effect to the Vision and Strategy.[36] The Vision and Strategy also forms part of the Waikato Regional Policy Statement.[37]

Additionally, the Vision and Strategy is an approved statement of general policy for various conservation-related enactments.[38] Management strategies or plans made under those enactments must not derogate from statements of general policy, thus including the Vision and Strategy. The Vision and Strategy will also prevail over any inconsistent provision of a bylaw made by a local authority, if it is more stringent than the bylaw.[39]

Finally, in carrying out functions or exercising powers, decision-makers under various enactments (including the *RMA*) 'must have particular regard to the vision and strategy' insofar as they relate to the Waikato River or activities in the catchment that affect the Waikato River.[40] The effect of the term 'have particular regard to' is noted below.

The paramount status of the Vision and Strategy has been highlighted because of the involvement of Māori in its development and subsequent acknowledgement of Māori interests in its terms.

The Vision and Strategy was developed by the Guardians Establishment Committee ('GEC'), set up under the 2007 Agreement in Principle.[41] The GEC was

31 *Waikato River Settlement Act* s 3.
32 Ibid s 5(2).
33 Ibid s 5(1).
34 Ibid s 12.
35 Ibid s 13.
36 Ibid s 15.
37 Ibid s 11.
38 Ibid s 16(2)–(6).
39 Ibid s 16(15).
40 Ibid s 17.
41 Waikato-Tainui and Her Majesty the Queen in right of New Zealand, *Agreement in Principle for the Settlement of the Historical Claims of Waikato-Tainui in relation to the Waikato River* (16 December 2007) cl 49.

the precursor to what was intended in the 2008 Deed of Settlement to be the more permanent 'Guardians of the Waikato River'.[42]

The GEC was comprised of 16 members, of which half were Māori.[43] The GEC prepared a consultation draft of the Vision and Strategy and undertook a public consultation process on the draft, after which it agreed upon a final version.[44] This is what is now found in schedule 2.

The influence of the Māori appointees to the GEC in the drafting and finalisation is evidenced in the final Vision and Strategy by the many objectives and strategies that refer to Māori concepts and aspirations.[45] In addition (and no doubt of symbolic significance), the opening statement of the Vision and Strategy is an excerpt from 'a famous lament by the second Māori King, Tāwhiao, in which he recorded his adoration for his ancestral river and the significance of the river as a treasure for all generations'.[46] It reads as follows:

> *Tooku awa koiora me oona pikonga he kura tangihia o te maataamuri.*
> [The river of life, each curve more beautiful than the last.][47]

This is followed by the vision for the Waikato River:

> Our vision is for a future where a healthy Waikato River sustains abundant life and prosperous communities who, in turn, are all responsible for restoring and protecting the health and wellbeing of the Waikato River, and all it embraces, for generations to come.[48]

As well as the Māori-oriented objectives, there are many other objectives and strategies contained in the Vision and Strategy, reflecting the input from the public consultation process and the other appointees, and the overarching purpose of the settlement. This overarching purpose has been prevalent from the first of the settlement documents[49] and is reflected in the first of the objectives in the Vision and Strategy.[50]

There was necessarily going to be Māori involvement and Māori interests acknowledged, given that the Vision and Strategy was developed in the context of

42 Her Majesty the Queen in right of New Zealand and Waikato-Tainui, *Deed of Settlement in Relation to the Waikato River* (22 August 2008) cls 6.1–6.9. This later body was amalgamated with the Waikato River Statutory Board in the 2009 Deed of Settlement to become the Waikato River Authority.

43 *Agreement in Principle*, above n 41, cl 50.

44 Ibid cls 51, 53.

45 See, eg, *Waikato River Settlement Act* sch 2 cls 1(3)(b), (c), (f); 2(b), (c), (f).

46 Linda Te Aho, 'Contemporary Issues in Māori Law and Society: The Tangled Web of *Treaty* Settlements – Emissions Trading, Central North Island Forests, and the Waikato River' (2008) 16 *Waikato Law Review* 229, 238.

47 *Waikato River Settlement Act* sch 2 cl 1(1).

48 Ibid sch 2 cl 1(2)

49 *Agreement in Principle*, above n 41, cl 37.

50 *Waikato River Settlement Act* sch 2 cl 1(3)(a).

a *Treaty* settlement. But the point to be made is that by virtue of the *Treaty* settlement process Māori were able to participate in the preparation of the initial draft, and its subsequent development and approval, they were not mere participants in the public consultation process on a draft already prepared.

Co-governance: the Waikato River Authority

The next major element of the settlement was the establishment of a co-governance entity for the Waikato River, the Waikato River Authority ('WRA'). The WRA emerged from the streamlining of the settlement that took place pursuant to the 2009 Deed of Settlement which combined several proposed statutory bodies into this single entity.[51]

The WRA has ten members, five of whom are appointed by various Māori trusts associated with the Waikato River, and five by the Minister.[52] So there is equal representation of Māori and the Crown on the WRA, reflecting the *Treaty* principle of partnership.

The purpose of the WRA is to:

> (a) set the primary direction through the vision and strategy to achieve the restoration and protection of the health and wellbeing of the Waikato River for future generations:
> (b) promote an integrated, holistic, and co-ordinated approach to the implementation of the vision and strategy and the management of the Waikato River:
> (c) fund rehabilitation initiatives for the Waikato River in its role as trustee for the Waikato River Clean-up Trust.[53]

Its general functions are set out in s 23, the principle function being to achieve its purpose. Importantly, though, the WRA has functions relating to resource consents. It is required to establish and maintain a register of accredited commissioners who are *RMA* decision-makers and have been appointed for inclusion on the register by Waikato-Tainui or *iwi* who appoint members of the WRA.[54] If an application is made to the Waikato Regional Council for a resource consent relating to the Waikato River[55] and the Council holds a hearing in relation to the application, then the hearing committee must consist of equal numbers of people appointed by the Council, and by the WRA from the register, along with an independent chairperson.[56] Therefore, assuming that the commissioners who have been appointed for inclusion on the register by Waikato-Tainui (and other *iwi* who appoint members to the WRA) are themselves Māori, this represents

51 Finlayson, above n 26.
52 *Waikato River Settlement Act* sch 6 cl 2.
53 Ibid s 22(2).
54 Ibid s 25.
55 Ibid s 26.
56 Ibid s 28.

considerable participation by Māori in resource consent approvals under the *RMA*. A similar mix of appointees is required for a board of inquiry established for call-ins under part 6AA of the *RMA*. [57]

Another of the WRA's functions is to review the Vision and Strategy to determine whether to make recommendations that it be amended. The importance of the Vision and Strategy as the primary direction-setting document for the Waikato River makes this another important function of the WRA.

In reviewing the Vision and Strategy, the WRA must take into account:

(i) the Waikato-Tainui environmental plan:
(ii) other iwi environmental plans to the extent to which they relate to the Waikato River:
(iii) the Waikato-Tainui objectives for the Waikato River:
(iv) other iwi objectives for the Waikato River:
(v) the report of the scoping study.[58]

The WRA may take into account any other documents it thinks are relevant to the health and wellbeing of the Waikato River;[59] however, Māori interests and objectives are clearly at the forefront of any review. Any recommendations made by the WRA must be approved by the Crown and the various Māori trust appointers before the Vision and Strategy can be amended.[60] So Māori are involved in the review (as members of the WRA) as well as in the subsequent approval of any amendments to the Vision and Strategy (as appointers to the WRA).

Finally, the WRA has an important role as the administrator of the Waikato River Clean-up Trust, for which it develops funding strategies.[61]

The inclusion of these functions for the WRA is a substantial improvement on the co-governance functions of the RLSG provided for in the *TALS Act*.

Co-management

The other aspect of the Waikato River Settlement seen as particularly significant was the extent of co-management arrangements. According to Ruru, '[t]hese are the first statutes in New Zealand to elevate Māori to co-management roles with the Crown in regard to fresh water'.[62]

57 Part 6AA relates to resource consents that are part of a proposal of national significance. Pursuant to this part, the Minister can call in a matter at his or her own initiative or at the request of an applicant or local authority. The WRA also has the power under s 23(2)(h) to request a call in.
58 *Waikato River Settlement Act* s 20(2)(a).
59 Ibid s 20(2)(b).
60 Ibid s 21(3).
61 See, eg, Waikato River Authority, *Waikato River Clean-up Trust Funding Strategy 2015*.
62 Ruru, 'Indigenous Restitution', above n 24, 334. Ruru includes here two other statutes that relate to the Waikato River: *Ngati Tūwharetoa, Raukawa, and Te Arawa*

The co-management arrangements involve the preparation of an Integrated River Management Plan,[63] a Waikato-Tainui Environment Plan[64] and joint management agreements ('JMAs').[65] The focus here will be on the JMA provisions because they are the most far-reaching of the co-management arrangements under the *Waikato River Settlement Act*, and are an example of Māori using the *Treaty* settlement process to improve on the under-utilised JMA provisions in the *RMA*.

JMAs under the *Waikato River Settlement Act* are intended to enhance collaborative planning for the Waikato River between local authorities and the Waikato Raupatu River Trust ('WRRT'), and are to be developed and implemented according to a number of guiding principles. These include the promotion of the overarching purpose of the settlement and the principle of co-management, a shared commitment to good faith, cooperation and transparency, and respect for the *mana whakahaera* rights and responsibilities of Waikato-Tanui.[66] Unlike JMAs under the *RMA*, JMAs under the *Waikato River Settlement Act* are mandatory.[67]

The scope of a JMA under the *Waikato River Settlement Act* is fairly wide; some content of the JMA is mandatory,[68] and some may be included by agreement.[69] The mandatory content includes monitoring and enforcement; preparation, review and amendments to *RMA* planning documents;[70] and the exercise of duties, functions or powers under part 6 of the *RMA* in relation to resource consent applications.[71]

With regard to resource consent applications, this does not require that the WRRT participate in the decision-making itself, but that the local authority and the WRRT work together in carrying out those duties, functions and powers. This includes timely provision of information to the WRRT about applications[72] and the joint development of decision-making criteria by the local authority and the WRRT.[73] However, such criteria are in addition to, and must not derogate from, the criteria that a consent authority must apply under the *RMA*, and they do not impose a requirement on the consent authority to change, cancel or review consent conditions.[74]

River Iwi Waikato River Act 2010 (NZ) and *Nga Wai o Maniapoto (Waipa River) Act 2012* (NZ).

63 *Waikato River Settlement Act* ss 35–38.

64 Ibid ss 39–40.

65 Ibid ss 41–55.

66 Ibid s 44.

67 Ibid s 41(1).

68 Matters referred to in s 43 and matters described in s 62 of the *Waikato River Settlement Act*.

69 Matters agreed under s 52 of the *Waikato River Settlement Act*.

70 Defined in s 6 of the *Waikato River Settlement Act*.

71 Ibid s 43.

72 Ibid s 47(2)(a)–(c).

73 Ibid s 47(2)(d).

74 Ibid s 47(3).

So the mandated role of the WRRT in relation to resource consent applications under the JMA provisions of the *Waikato River Settlement Act* is not as strong as the role of the WRA, as the WRA can appoint people to hear and determine applications for resource consents (ie participate in making the decision) in circumstances where the consent authority decides that a hearing is required.[75] The WRRT and a local authority may, however, extend the JMA to cover 'any other duties, functions or powers they agree on',[76] which could conceivably include a greater role for the WRRT in decision-making on resource consent applications. But this suffers from a similar problem as JMAs under the *RMA* in that local authorities are generally reluctant to share power. On a positive note, with a mandated JMA, local authorities are required to work closely with the WRRT, thus building relationships and increasing capacity, two factors noted as inhibiting the take-up of JMAs under the *RMA*. [77]

Pursuant to the *Waikato River Settlement Act*, the Waikato Regional Council entered into a JMA in 2013 with Waikato-Tainui Te Kauhanganui Inc, in its capacity as trustee of the WRRT. The JMA affirms the commitment of the parties to

> recognise [that] the relationship underpinning this Agreement is an evolving one whereby Waikato-Tainui and the Council will continue to explore the potential for transfers, delegations and sharing of relevant powers and functions under the *Resource Management Act 1991* (RMA) and *Local Government Act 2002* (LGA 2002).[78]

So this JMA contemplates the possibility of a greater role for Māori in decision-making under the *RMA*.

The local authority and the Trust can agree to suspend (in whole or in part) the operation of the JMA,[79] but, in contrast to JMAs under the *RMA*, neither party has the right to terminate the JMA.[80] Thus there is more security for the parties in the longevity of a JMA made under the *Waikato River Settlement Act*.

There is one similarity with JMAs made under the *RMA*, namely that where a statutory function or power is affected by a JMA, the local authority can, in limited circumstances, exercise that function or power on its own account, and not in accordance with the JMA.[81]

75 *RMA* s 100.
76 *Waikato River Settlement Act* s 52(1).
77 Elizabeth Clark, 'Section 33 of the *Resource Management Act 1991*' in Janine Hayward, *Local Government and the Treaty of Waitangi* (Oxford University Press, 2003) 49–50.
78 Waikato Raupatu River Trust and Waikato Regional Council, *Joint Management Agreement* (18 June 2013) cl 1.5(d).
79 *Waikato River Settlement Act* s 49.
80 Ibid s 51(5). If, however, the JMA is extended to include other duties, functions or powers, the extended part can be terminated by one party giving notice to the other, provided certain conditions are met: ibid s 52(4), (5).
81 Ibid s 55.

The JMA provisions of the *RMA* do not apply to the making of a JMA under the *Waikato River Settlement Act*;[82] however, that does not prevent the WRRT and the local authority from making other JMAs under the *RMA*, or co-management agreements under any enactment.[83] But given that a JMA under the *Waikato River Settlement Act* can be extended to include additional matters not mandated by that Act, it seems unlikely that the WRRT and a local authority would seek to utilise the JMA provisions of the *RMA*, rather than simply extend their JMA under the *Waikato River Settlement Act.* [84]

The Waikato River JMA provisions are clearly more progressive than the JMA provisions in the *RMA*, and overcome some of the *RMA's* deficiencies.

Concluding remarks on the Waikato River Settlement

Compared with the two earlier river settlements outlined in this chapter, the Waikato River Settlement broke new ground in relation to the participation of Māori in the management of water resources, introducing a number of new elements to achieve this goal. It has also gone some way to addressing some of the deficiencies in the *RMA* for achieving Māori aspirations for river management. Co-governance by way of equal representation on the WRA has given Māori an equal say in certain matters concerning the Waikato River, not least being the WRA's ability to appoint commissioners to conduct hearings on resource consent applications and its role in reviewing the Vision and Strategy, the primary direction-setting document for the Waikato River. In addition, mandated and enduring JMAs require local authorities and the WRRT to work together to improve Māori engagement in matters relating to the management of the Waikato River.

It remains to consider whether these 'bold' and 'groundbreaking' developments in relation to the Waikato River might have any application outside of Aotearoa New Zealand, a task which will be undertaken after an analysis of the Whanganui River Settlement.

The Whanganui River Settlement

Introduction

On 30 August 2012 the New Zealand Government announced that it had reached a framework agreement with the negotiators for the Whanganui Iwi for

82 Ibid s 51(1).

83 Ibid s 54.

84 Note that pursuant to s 80 of the *Waikato River Settlement Act*, the Trust and the local authority are required to enter into a co-management agreement in relation to the fee simple sites, managed properties and reserved sites vested in the Trust pursuant to the settlement. Such an agreement was entered into in 2012: Waikato Raupatu River Trust and Waikato Regional Council, *Co-management Agreement for Waikato River Related Lands* (10 December 2012).

the settlement of its long-running claim to the Whanganui River.[85] In what was seen as an innovative development in water management in not only Aotearoa New Zealand but internationally, included in this framework was an in-principle agreement to give legal personality to the Whanganui River, the embodiment of that legal personality being in the form of a river guardian.

The concept of giving legal personality to a natural object has existed in theory since Christopher Stone's seminal article of 1972, 'Should Trees Have Standing?'[86] The basic idea behind Stone's concept is that many inanimate entities such as corporations and trusts have legal personality which gives them legal rights,[87] so why not extend this to natural objects, such as trees and rivers? He suggests that those rights can be protected by the appointment of a guardian, who can then represent the natural object in court proceedings (standing).[88] Stone's concept has started to gain traction in recent years; in 2008 the rights of nature, or 'Pacha Mama', were recognised in Ecuador's *Constitution.* [89] In 2010 the World People's Conference on Climate Change and the Rights of Mother Earth adopted the *Universal Declaration on the Rights of Mother Earth.* [90] And in 2012 the International Union for the Conservation of Nature adopted a resolution calling for the development of a Universal Declaration of the Rights of Nature.[91] More recently, the Ganga and Yamuna Rivers in India and the Atrato River in Columbia have been recognised as legal persons.[92]

But Aotearoa New Zealand was the first country to give legal personality to a specific natural object.[93] This has ostensibly introduced a new type of governance structure for natural objects in Aotearoa New Zealand, one that is clearly focussed

85 Christopher Finlayson, 'Whanganui River Agreement Signed' (Press Release, 30 August 2012).

86 Christopher Stone, 'Should Trees Have Standing? Towards Legal Rights for Natural Objects' (1972) 45 *Southern California Law Review* 450.

87 Ibid 452.

88 Ibid 464 ff.

89 *República del Ecuador Constitucion de 2008* [Republic of Ecuador Constitution of 2008] arts 10, 71–4, Political Database of the Americas <http://pdba.georgetown.edu/Constitutions/Ecuador/english08.html>.

90 *Universal Declaration on the Rights of Mother Earth*, World People's Conference on Climate Change and the Rights of Mother Earth (22 April 2010). Global Alliance for the Rights of Nature <http://therightsofnature.org/universal-declaration>.

91 International Union for the Conservation of Nature, *Incorporation of the Rights of Nature as the Organisational Focal Point in IUCN's Decision Making*, Res 100, World Conservation Congress (6–15 September 2012).

92 Erin O'Donnell and Julia Talbot-Jones, 'Legal Rights for Rivers: What Does This Actually Mean?' (2017) *Australian Environment Review* 159, 159.

93 The potential application in Aotearoa New Zealand of the concept had been advocated in the early 1990s by Alex Frame: Alex Frame, 'Property and the *Treaty of Waitangi*: A Tragedy of the Commodities' in Janet McLean, *Property and the Constitution* (Hart, 1999) 237. See also James D K Morris and Jacinta Ruru, 'Giving Voice to Rivers: Legal Personality as a Vehicle for Recognising Indigenous Peoples' Relationships to Water?' (2010) 14 *Australian Indigenous Law Reporter* 49, 56. The concept was first applied in relation to the Te Urewera National Park: *Te Urewera Act 2014* (NZ).

on the environment, but an environment which is shaped by and reflective of Māori concepts and values. In that respect, this particular adaptation of Stone's original concept takes a much more holistic approach to environmental protection, one which acknowledges the intrinsic relationship which Indigenous people have with the environment.

Background

The Whanganui River is Aotearoa New Zealand's longest navigable river and is of significant national importance.[94] Its importance in particular to the Whanganui Iwi is reflected in their ongoing struggle for well over a century to have their rights and interests in the Whanganui River recognised,[95] including one of the longest running legal battles in Aotearoa New Zealand's history.[96] Despite losing this legal battle, the Whanganui Iwi continued to pursue various legal proceedings, eventually lodging a claim before the Waitangi Tribunal in 1990.[97] The Tribunal produced its final report on the river in 1999 in which it recommended various options for consideration in future negotiations, none of which were explicitly adopted in the settlement. But there can be little doubt that the content of the report and the findings of the Tribunal were influential.[98]

Negotiations between the Whanganui Iwi and the Crown to settle the Whanganui Iwi's historical *Treaty* claim to the river commenced in 2002.[99] Although those negotiations ended unsuccessfully, by 2011 the parties had signed a Record of Understanding, in which they 'agreed to enter into formal negotiations to settle the historical *Treaty* claims of Whanganui Iwi in relation to the Whanganui River'.[100] Following the signing of an in-principle framework agreement in August 2012, negotiations continued and on 26 March 2014 the New Zealand Government announced that a Deed of Settlement had been initialled by the negotiators.[101] After

94 Whanganui Iwi and the Crown, *Ruruku Whakatupua – Te Mana o Te Iwi o Whanganui* (5 August 2014) 6 ('*Ruruku Whakatupua – Te Mana o Te Iwi o Whanganui*') <www.wrmtb.co.nz/new_updates/RurukuWhakatupua-TeManaoTeIwioWhanganui.pdf>.

95 Formal objections began in 1873 with a parliamentary petition against the Timber Floating Bill: Waitangi Tribunal, *Whanganui River Report* (1999) 4 ('*Whanganui River Report*').

96 An extensive discussion of the legal battle is contained in the *Whanganui River Report*, above n 95, 195–232.

97 Ibid 357.

98 The in-principle framework agreement between the Whanganui Iwi and the Crown refers to various findings of the Tribunal: Whanganui Iwi and the Crown, *Tūtohu Whakatupua* (30 August 2012) <www.wrmtb.co.nz/new_updates/TuutohuWhakatupuaFinalSigned.pdf>.

99 Whanganui Iwi and the Crown, *Record of Understanding* (13 October 2011) [1.16] <www.wrmtb.co.nz/new_updates/Record%20of%20Understanding%202012.pdf>.

100 Ibid [1.25].

101 Christopher Finlayson, 'Whanganui River Deed of Settlement Initialled' (Media Release, 26 March 2014) <www.beehive.govt.nz/release/whanganui-river-deed-settlement-initialled>.

ratification by Whanganui Iwi members,[102] the Deed of Settlement was signed on 5 August 2014.[103] The Deed of Settlement comprised two documents. The first, *Ruruku Whakatupua – Te Mana o Te Awa Tupua*, [104] related to the recognition of the Whanganui River as a legal entity. The second, *Ruruku Whakatupua – Te Mana o Te Iwi o Whanganui*, [105] contained all the other elements of the settlement. On 2 May 2016 the Te Awa Tupua (Whanganui River Claims Settlement) Bill 2016 (NZ) was introduced into the New Zealand Parliament, finally becoming law in March 2017.

Summary of the settlement

The *Te Awa Tupua (Whanganui River Claims Settlement) Act 2017* (NZ) ('*Te Awa Tupua Act*') provides for the recognition of the Whanganui River, Te Awa Tupua, as a legal person, pursuant to what has been described here and by other commentators as a guardianship model,[106] being the model espoused by Stone. But where this approach differs from Stone's model is that it incorporates a distinctively Māori worldview, a view in which Māori see rivers as the embodiment of their ancestors, 'tupuna'.[107] Māori terminology is therefore used throughout the settlement documents and in the *Te Awa Tupua Act*.

Under the Whanganui River model, the guardian of Te Awa Tupua is Te Pou Tupua – 'the human face of Te Awa Tupua'.[108] Te Pou Tupua comprises two people, one nominated by the Crown and one nominated by the *iwi* with interests in the Whanganui River.[109] Once appointed to Te Pou Tupua, they act collectively on behalf of Te Awa Tupua, not on behalf of their nominators.[110]

102 Whanganui River Māori Trust Board, *Whanganui River Settlement – Ratification Results* <www.wrmtb.co.nz/Ratification%20Results%20Panui.pdf>.
103 Christopher Finlayson, 'Whanganui River Deed of Settlement Signed' (Media Release, 5 August 2014).
104 Whanganui Iwi and the Crown, *Ruruku Whakatupua – Te Mana o Te Awa Tupua* (5 August 2014) <www.wrmtb.co.nz/new_updates/Ruruku_Whakatupua_Te_Mana_o_Te_Awa_Tupua_Signed5August%202014.pdf> ('*Ruru Whakatupua – Te Mana o Te Awa Tupua*').
105 *Ruruku Whakatupua – Te Mana o Te Iwi o Whanganui*, above n 94.
106 Ruru, 'Indigenous Restitution', above n 24, 341; Meg Good, 'The River as a Legal Person: Evaluating Nature Rights-Based Approaches to Environmental Law' (2013) 1 *National Environmental Law Review* 34, 35; Tom Barraclough, *How Far Can the Te Awa Tupua (Whanganui River) Proposal Be Said to Reflect the Rights of Nature in New Zealand?* (LLB(Hons) Thesis, University of Otago, 2013) 23.
107 For more on the Māori world view of rivers, see generally *Whanganui River Report*, above n 95, 36 ff.
108 *Te Awa Tupua Act 2017* s 18(2).
109 Ibid s 20(1)–(2).
110 Ibid s 19(2)(a).

In undertaking its functions,[111] Te Pou Tupua must uphold the four intrinsic values of Te Awa Tupua,[112] which are known as Tupua te Kawa.[113] It is in Tupua te Kawa where one finds the embodiment of the Māori worldview.

Any person exercising functions under 25 named statutes[114] 'must recognise and provide for' the Te Awa Tupua status as a legal person, and for Tupua te Kawa.[115] Three statutes are singled out for differential treatment: the *Heritage New Zealand Pouhere Taonga Act 2014* (NZ), the *Public Works Act 1981* (NZ) and the *RMA*. [116] Any person exercising powers under these three Acts must 'have particular regard to' the Te Awa Tupua status and Tupua te Kawa.[117] '[R]ecognise and provide for' has been interpreted by the courts as being stronger than the phrase 'have particular regard to'.[118] Therefore, this means that there is a lesser standard required of decision-makers under these three Acts.

One function of Te Pou Tupua is to enter into relationship documents with Crown agencies and local authorities concerning various matters of mutual interest, including (in relation to local authorities) the exercise of functions and powers in relation to the granting of consents relating to the Whanganui River.[119] It is also assisted by an advisory group of three, known as Te Karewao, of which two are Māori.[120]

An important element of the legal framework for Te Awa Tupua is the preparation of the Te Awa Tupua Strategy – Te Heke Ngahuru ki Te Awa Tupua ('Te Heke Ngahuru') to identify issues relating to the Whanganui River, provide a strategy to address those issues and recommend actions to be taken.[121] Te Heke Ngahuru will be prepared by a strategy group, known as Te Kōpuka nā Te Awa Tupua ('Te Kōpuka').[122] Because the purpose of Te Kōpuka is intended to be collaborative,[123] the group will be relatively large, comprising up to 17 members, one appointed by the trustees (of Ngā Tāngata Tiaki o Whanganui) and up to five members appointed by the *iwi* with interests in the Whanganui River.[124] Further,

111 Ibid s 19(1).
112 Ibid s 19(1)(b)(ii), (2)(a).
113 Ibid s 13.
114 Ibid sch 2 cl 1.
115 Ibid s 15(2).
116 Insofar as the *RMA* does not apply to preparing or changing a regional policy statement, regional plan or district plan made under the *RMA*, which are covered in s 15 (2) and therefore must be recognised and provided for.
117 *Te Awa Tupua Act* s 15(3).
118 *Takamore Trustees v Kapiti Coast District Council* [2003] 3 NZLR 496, [89].
119 *Te Awa Tupua Act* s 19(1)(h). The details are set out in *Ruruku Whakatupua – Te Mana o Te Awa Tupua*, above n 104, cls 3.36–3.42.
120 *Te Awa Tupua Act* s 28(1). An additional member can be appointed to represent specific *iwi* interests where Te Pou Tupua is exercising a function relating to a discrete part of the River: s 28(2).
121 Ibid s 36.
122 Ibid s 30(1).
123 Ibid s 29(3).
124 Ibid s 32(1). The remaining appointees are a combination of local authorities, various interest groups and an appointee from the Director General of Conservation.

in performing its functions, Te Kōpuka must 'have particular regard to' the Te Awa Tupua status and Tupua te Kawa.[125]

The legal effect of Te Heke Ngahuru is that any person exercising functions, duties or powers under any of the 26 named statutes must 'have particular regard to' it.[126] As mentioned above, this is a lesser requirement on decision-makers than 'recognise and provide for', and therefore the implementation of its recommended actions is not inevitable. However, if one considers the role of regional and district plans in relation to applications for resource consents under the *RMA*, Te Heke Ngahuru is in a better position than these documents, to which decision-makers are only required to 'have regard'. The *Te Awa Tupua Act* also provides that decision-makers may adopt (in whole or in part) Te Heke Ngahuru as part of an *RMA* planning document, namely a regional policy statement, regional plan or district plan.[127] Accordingly, decision-makers must have particular regard to Te Heke Ngahuru as a stand-alone document, but may also be required to have regard to it, if it is adopted as part of an *RMA* planning document. It is not, however, as important in the *RMA* hierarchy as, for example, the Waikato River Vision and Strategy, because it does not prevail over any inconsistent provisions in a planning or policy document issued under the *RMA*, [128] nor does the Act mandate that *RMA* planning documents be amended to conform with Te Heke Ngahuru.[129] It is merely discretionary.

The *Te Awa Tupua Act* also vests in Te Awa Tupua that part of the bed of the Whanganui River owned by the Crown,[130] and establishes a fund to support the health and wellbeing of Te Awa Tupua.[131] Te Pou Tupua also maintains a registry of hearing commissioners for the purposes of resource consent applications relating to the Whanganui River.[132]

Other arrangements include the protection of the name of Te Awa Tupua,[133] the deeming of Te Awa Tupua as a public authority under the *RMA*, [134] and as a body corporate so that it can make an application to be a heritage protection authority ('HPA') under the *RMA*. [135]

The *Te Awa Tupua Act*, however, does not create, limit, transfer, extinguish or otherwise affect any rights to, or interests in, water.[136] Nor does the vesting of the Crown-owned parts of the bed of the Whanganui River in Te Awa Tupua create

125 Ibid s 30(3).
126 Ibid s 37(1)–(2). Note that the 26 statutes include the *RMA* and the *Heritage New Zealand Pouhere Taonga Act 2014* (NZ). It would appear, however, that the *Public Works Act 1981* (NZ) is exempt.
127 *Te Awa Tupua Act* s 37(5)–(6).
128 *Waikato River Settlement Act* s 12(1).
129 *Te Awa Tupua Act* s 38.
130 Ibid s 41.
131 Ibid s 57.
132 Ibid ss 19(1)(f), 61(3).
133 Ibid s 60.
134 Ibid s 17(e).
135 Ibid s 17(f).
136 Ibid s 16(b).

or transfer a proprietary interest in water.[137] Further, the consent of Te Pou Tupua is not required to use water, although a consent authority may determine that Te Pou Tupua is an affected person for the purpose of applications for resource consents relating to water, which gives it certain procedural rights under the *RMA*. [138]

Finally, existing private property rights will not be affected by the vesting of the Crown-owned parts of the bed of the Whanganui River in Te Awa Tupua,[139] nor will customary rights or title.[140]

Focussing on what has been described above, this chapter engages first with the merits of the settlement for the Whanganui Iwi (and, by extension, for Māori more generally), as subsequently reflected in the *Te Awa Tupua Act*, and then analyses some of its deficiencies in that respect.

Critique of the settlement and the river as a legal entity

Merits of the settlement

RECOGNITION AS A SINGLE ENTITY

The *Te Awa Tupua Act* recognises the River, Te Awa Tupua as 'an indivisible and living whole … from the mountains to the sea'.[141] This not only reflects the Māori view of the river, but recognises the value in having a unified approach to protecting the health and wellbeing of Te Awa Tupua. This unified approach is further evidenced by the requirement for Te Kōpuka, the strategy group, to develop *Te Heke Ngahuru*, the river strategy.

LEGAL PERSONALITY AND STANDING

Stone emphasised standing (the ability to bring a court action) as a vital element of natural objects being granted legal personality. In the context of the Whanganui River Settlement, the status of Te Awa Tupua as a legal entity means that people may be held accountable for damage to the Whanganui River without an individual having to show that their particular interests have been affected. If the river is damaged in some way (damage in this respect refers to damage that has not been authorised by a licence or permit) and if any of the Tupua te Kawa are affected, then Te Pou Tupua would be able to bring an action on behalf of Te Awa Tupua.[142] Standing is effectively guaranteed, even if the outcome of

137 Ibid s 46(1)(a).
138 An affected person is determined pursuant to s 95E of the *RMA*. This entitles them to be notified about and put in submissions about resource consent applications for which only limited notification has been given: *RMA* ss 95B(2), 96(3).
139 *Te Awa Tupua Act* ss 16(a), 46(2)(b).
140 Ibid s 46(2)(b).
141 Ibid ss 12, 13(b).
142 Ibid s 14(2).

proceedings is not. This an important procedural benefit of the settlement as it ensures access to the courts, and in doing so it emphasises Māori values as a basis for bringing a court action.

THE RMA AND WATER MANAGEMENT

Te Awa Tupua, via Te Pou Tupua (with its Māori appointee), has the potential to have decision-making powers over itself. Pursuant to the *Te Awa Tupua Act*, it is deemed to be a public authority for the purposes of the *RMA*. [143] This means that a local authority is able to transfer one or more of its powers or functions to Te Awa Tupua.[144] The most important function of a local authority in this regard is its role as a consent authority.[145] Accordingly, despite the settlement not including a provision deeming Te Awa Tupua to be a consent authority for the purposes of the *RMA*, the deeming of Te Awa Tupua as a public authority means that it is a possibility. As a consent authority, Te Awa Tupua's permission would be required to carry out any activity for which a resource consent is required under the *RMA*. In addition, being deemed a public authority also means that Te Awa Tupua can enter into JMAs with local authorities pursuant to section 36B of the *RMA*. Although these sections of the *RMA* have been little utilised to date,[146] this may change over time. The *Te Awa Tupua Act* provides for the making of relationship agreements with local authorities concerning (among other things) the exercise of functions and powers in relation to the granting of consents relating to the Whanganui River and the relationship between the exercise of the local authority's functions and the functions of Te Pou Tupua.[147] This would appear to be the forum for Te Pou Tupua to raise the issue of the transfer of functions under section 33 of the *RMA* from the local authority to Te Awa Tupua, represented by Te Pou Tupua. Te Pou Tupua, with one appointee nominated by the Crown, may be perceived as more neutral than an *iwi* authority, and technical expertise and resources may also be less of an issue. Thus, there might be less reluctance from local authorities to transfer functions to or enter into a JMA with Te Awa Tupua.

While the heritage protection provisions of the *RMA* are understood as having been of little utility in facilitating Māori participation in water resource management, this was largely because of the difficulties in being approved as

143 Ibid s 17(e).

144 *RMA* s 33.

145 Section 2 of the *RMA* defines a consent authority as 'a regional council, a territorial authority, or a local authority that is both a regional council and a territorial authority, whose permission is required to carry out an activity for which a resource consent is required under this Act'.

146 David V Williams, 'Ko Aotearoa Tēnei: Law and Policy Affecting Māori Culture and Identity' (2013) 20 *International Journal of Cultural Property* 311, 320; Waitangi Tribunal, *Ko Aotearoa Tēnei: A Report into Claims Concerning New Zealand Law and Policy Affecting Māori Culture and Identity – Te Taumata Tuarua* (Report of Wai 262 Tribunal, 2011) Vol 1, 113–14; Morris and Ruru, above n 93, 51.

147 *Te Awa Tupua Act* s 19(h).

an HPA.[148] However, Te Awa Tupua is deemed to be a body corporate specifically for the purpose of enabling it to apply to be an HPA.[149] This suggests that Te Awa Tupua would have little difficulty in being approved as an HPA for the Whanganui River. It is clearly appropriate for the role, and with the settlement providing financial support to Te Pou Tupua,[150] it should be able to carry out any financial responsibilities associated with being an HPA.[151] The benefits of being an HPA are, in summary, that it would potentially enable Te Awa Tupua to have a specific management function of a protective nature, in relation to at least that part of itself which is the subject of a heritage order.

RELEVANCE AND LEGAL EFFECT OF TE AWA TUPUA AND TUPUA TE KAWA

The entire legal framework, *Te Pā Auroa nā Te Awa Tupua*, with its focus on Māori values, is a relevant consideration for all statutory functions, powers and duties relating to the Whanganui River or to activities occurring in its catchment.[152] This is an important procedural benefit because it ensures that decision-makers always consider the Te Awa Tupua legal framework; it does not rely on a public official having to make a judgment call about whether or not the framework is relevant.

In addition, any person in exercising functions, duties or powers under 25 named statutes (including certain parts of the *RMA*), must 'recognise and provide for' the Te Awa Tupua status and Tupua te Kawa.[153] Therefore, in these 25 statutes Te Awa Tupua and its values are more than just relevant considerations. They must be recognised and provided for, thus ensuring that they are reflected in the outcome, rather than simply considered in the process leading to the outcome. This arguably amounts to a substantive benefit, rather than a procedural one. Support for this view can be found in *Bleakley v Environmental Risk Management Authority* [154] in relation to the *Hazardous Substances and New Organisms Act 1996* (NZ), a view which was subsequently accepted in relation to the *RMA* by the Environment Court in *Ngati Hokopu Ki Hokowhitu v Whakatane District Council.* [155]

However, in the High Court decision of *Takamore Trustees v Kapiti Coast District Council* ('*Takamore Trustees*'), Ronald Young J in considering the difference in terminology between sections 6, 7 and 8 of the *RMA*, did not go quite so

148 Katie O'Bryan, *From Aqua Nullius to Aqua Minimus? The Legal Recognition in Victoria of Indigenous Rights to Participate in the Management of Inland Water Resources* (PhD Thesis, Monash University, 2015) 309–11.

149 *Te Awa Tupua Act* s 17(f).

150 *Ruruku Whakatupua – Te Mana o Te Awa Tupua*, above n 104, cls 3.31–3.35. Funding arrangements have not been included in the *Te Awa Tupua Act.*

151 These are the two matters which the Minister must consider when approving an application to become an HPA: *RMA* s 188(5).

152 *Te Awa Tupua Act* s 11(1).

153 Ibid s 15(2). In relation to the *RMA*, this applies to the preparation or changing of regional policy statements, regional plans and district plans: Ibid sch 2 cl 1(s).

154 [2001] 3 NZLR 213 [72].

155 [2002] 9 ELRNZ 111 [36].

far as to suggest that 'recognise and provide for' amounted to a substantive right.[156] The reasoning of Ronald Young J is somewhat confusing,[157] but what is clear from these cases is that 'recognise and provide for' is, at the least, considered to be a very strong directive which must be complied with.

In relation to the *Heritage New Zealand Pouhere Taonga Act 2014* (NZ), the *Public Works Act 1981* (NZ) and parts of the *RMA*, decision-makers must 'have particular regard to' the Te Awa Tupua status and Tupua te Kawa. This too is a procedural benefit and according to Ronald Young J, is less firm than 'recognise and provide for'.[158] Nonetheless, it is still a relatively strong directive and would certainly elevate Te Awa Tupua and Tupua te Kawa above other matters to which a person must only 'have regard' when exercising functions, duties or powers under those three Acts.[159]

COMPOSITION OF TE KŌPUKA

Te Kōpuka contains a substantial proportion of *iwi* members, with a maximum of six out of 17 able to represent *iwi* interests.[160] *Iwi* therefore form the largest representative grouping in Te Kōpuka.[161] Thus, assuming they work together as a bloc, *iwi* will have the strongest voice in Te Kōpuka. Although the Māori voice is still a minority one, in comparison with the level of mandated Indigenous representation on water-related committees in Australia, Māori representation is significantly higher. For example, the entity which most closely resembles Te Kōpuka in Victoria, the Birrarung Council (which will assist in the preparation of a strategic plan for the Yarra River), mandates a minimum of two Indigenous representatives out of 12.[162] Another example is the Basin Community Committee (established under the *Water Act 2007* (Cth) to provide advice to the Murray–Darling Basin Authority) which makes provision for up to two Indigenous representatives on the committee out of a total of up to 17 members.[163]

PROMOTES RELATIONSHIPS AND PRESERVES RIGHTS

Another positive element of the settlement is that it allows for the making of relationship agreements between Te Pou Tupua and local authorities.[164] Although

156 [2003] 3 NZLR 496, [89].
157 He uses 'take into account' in relation to all three sections: ibid.
158 Ibid.
159 In relation to the *RMA*, it would even seem to elevate them above the Principles of the *Treaty of Waitangi*, which decision makers 'shall take into account': *RMA* s 8. But it would not elevate them above any of those matters in s 7, as these are all matters to which decision makers must also 'have particular regard'.
160 *Te Awa Tupua Act* s 32(1)(a)–(b).
161 Government representatives make up the next largest bloc, being up to four local authority representatives, and one representative appointed by the Director-General of Conservation.
162 *Yarra River Protection (Wilip-gin Birrarung murron) Act 2017* (Vic) s 49(1)(a).
163 *Water Act 2007* (Cth) s 202(4)–(5).
164 *Te Awa Tupua Act* s 19(1)(h).

it does not mandate that agreement be reached, it is, nonetheless, a positive statement of intent which at least compels the parties to discuss relevant issues, and, in the context of the rest of the settlement, could well lead to an agreement between Te Pou Tupua and the relevant local authorities.

The *Te Awa Tupua Act* also preserves any existing customary rights and title,[165] as well as rights to apply for and be granted a customary rights order in relation to the Whanganui River,[166] the trade-off being that all other existing rights are also preserved.[167] This is important because following *Attorney-General v Ngati Apa*,[168] the continuing existence of Māori customary title over water is yet to be resolved.[169]

The settlement also settles only that part of the Whanganui Iwi's historical *Treaty* claim that relates to the Whanganui River.[170] This means that the Whanganui Iwi can continue to pursue its historical *Treaty* claim in respect of the remainder of the claim that does not involve the Whanganui River. It also means that contemporary actions for any breaches of the *Treaty* can still be brought against the Crown.

PROMINENCE OF MĀORI LANGUAGE AND VALUES

Finally, Māori language and values are very prominent throughout the *Te Awa Tupua Act*, a point which was noted in relation to the settlement documents preceding the Act.[171]

For example, the *Te Awa Tupua Act* sets out the four intrinsic values of Te Awa Tupua to be protected, Tupua te Kawa, which are clearly centred around Māori values. They recognise that the Whanganui Iwi and the River are interdependent, which is reflected particularly in the third of the four values: *Ko au te Awa, ko te Awa ko au* [I am the River and the River is me].[172] Headings to each part in the Act have a Māori component written first, followed by an English component.

With such prominence given to Māori language and values, this could have a further practical effect in that Māori perspectives on water management will be given greater priority than may otherwise be the case.

165 Ibid s 46(2)(b).
166 Ibid ss 80, 81(2).
167 Ibid ss 16(a), 46(2).
168 [2003] 3 NZLR 643.
169 Jacinta Ruru, 'Undefined and Unresolved: Exploring Indigenous Rights in Aotearoa's Freshwater Legal Regime' (2009) 20 *Journal of Water Law* 236, 241–2.
170 *Te Awa Tupua Act* s 9 contains the definition of historical claims, which excludes claims by the Whanganui Iwi that do not relate to the Whanganui River: s 9(4). Section 87 of the Act then settles all historical claims as defined.
171 Linda Te Aho, '*Ruruku Whakatupua Te Mana o te Awa Tupua* — Upholding the Mana of the Whanganui River' (2014) *Māori Law Review* ('Upholding the Mana of the Whanganui River').
172 *Te Awa Tupua Act* s 13(c).

Deficiencies of the settlement

The granting of legal personality to the Whanganui River does, however, raise some important issues for the Whanganui Iwi.

MANAGEMENT ISSUES

Recognition of Te Awa Tupua as a legal entity means that the Whanganui Iwi is a step removed from direct involvement in Te Awa Tupua. *Iwi* with interests in the Whanganui River nominate one of the two members who comprise Te Pou Tupua, but once appointed, that nominee does not represent those *iwi.* [173] In other words, Te Pou Tupua does not represent the interests of Māori; it represents the interests of the river, Te Awa Tupua. This is alleviated somewhat by a requirement that Te Pou Tupua

> develop appropriate mechanisms for engaging with, and reporting to, the iwi and hapū with interests in the Whanganui River on matters relating to Te Awa Tupua, as a means of recognising the inalienable connection of those iwi and hapū with Te Awa Tupua.[174]

It must also be remembered that the values of Te Awa Tupua to be protected, Tupua te Kawa, were negotiated with the Whanganui Iwi. These values recognise the relationship and role of the Whanganui Iwi with regard to Te Awa Tupua.

Te Awa Tupua, through its guardian, Te Pou Tupua, still has a relatively limited role in the management of the River. Te Pou Tupua is not a consent authority. It does not prepare or assist in the preparation of Te Heke Ngahuru (the strategy) nor is it represented on Te Kōpuka (the strategy group). Te Pou Tupua has the status of a landowner in relation to the beds of those parts of the river which have been vested in Te Awa Tupua, which does not give it a management role. It does, however, administer and make decisions regarding applications to the fund that has been set up to support the health and wellbeing of Te Awa Tupa.[175] Te Awa Tupua also has the *potential* to take on some management responsibilities via the *RMA* under a s 33 transfer or a s 36B JMA. It can also apply to become an HPA. If any of this were to occur, Te Pou Tupua, its human face, would exercise those management responsibilities on its behalf. But these management roles are not as of right.

In addition, although Te Awa Tupua is required to maintain a register of hearing commissioners for resource consent applications relating to the Whanganui River,[176] it does not make appointments; that role is undertaken by the relevant authority (albeit in consultation with Te Pou Tupua).[177] Furthermore,

173 Ibid s 19(2)(a).
174 Ibid s 19(2)(b).
175 Ibid ss 19(1)(e), 58.
176 Ibid s 19(1)(f).
177 Ibid sch 6 cl 5(3)(b).

appointments from the register by the relevant authority are not a mandatory requirement.[178]

Accordingly, there is effectively little change to the current governance of the Whanganui River. Any management role is contingent on an agreement to transfer powers, entry into a JMA or a successful application to become an HPA, all of which occur under the *RMA*.

NO RECOGNITION OF OWNERSHIP OF WATER

The settlement has been criticised for the lack of recognition that it gives to ownership of water by Te Awa Tupua.[179] Thus, from a practical perspective, the recognition of Te Awa Tupua as 'an indivisible and living whole comprising the Whanganui River from the mountains to the sea, incorporating all its physical and metaphysical elements'[180] is largely illusory, of only symbolic effect. Not only is water an essential physical element of Te Awa Tupua, but water's metaphysical elements are what bind it to the rest of Te Awa Tupua.

FRAGMENTATION OF OWNERSHIP OF THE RIVER BED

The illusory nature of Te Awa Tupua as 'an indivisible and living whole' is further exacerbated by the fact that it is only those parts of the river bed which are owned by the Crown that are being transferred in title to Te Pou Tupua.[181] It also excludes any part of the bed of the Whanganui River which is located in the marine or coastal area.[182] Thus, despite the intentions expressed in the settlement, Te Awa Tupua is still physically (and therefore metaphysically) fragmented. This, however, is less likely to be an issue in the Victorian context, as the Crown in Victoria owns the beds and banks of all major rivers.[183]

NO GUARANTEE OF PROTECTION

Another criticism is that '[legal] personhood and standing *in themselves* confer nothing except procedural access and capability'.[184] In other words, Te Pou Tupua on behalf of Te Awa Tupua will have standing to bring an action for damage to Te Awa Tupua, but this does not guarantee that the action will succeed.[185]

178 Ibid sch 6 cl 5(1)(b).
179 Te Aho, 'Upholding the Mana of the Whanganui River', above n 171; Laura Hardcastle, 'Turbulent Times: Speculations about How the Whanganui River's Position as a Legal Entity Will Be Implemented and How It May Erode the New Zealand Legal Landscape' (February 2014) *Māori Law Review* 4.
180 *Te Awa Tupua Act* s 12.
181 Ibid s 41(1).
182 Ibid s 41(2)(d)(ii).
183 This occurred by virtue of s 5 of the *Water Act 1915* (Vic).
184 Barraclough, above n 106, 47.
185 Hardcastle suggests that causation could be a problem: Hardcastle, above n 179, 7–8.

As acknowledged earlier, procedural access does not necessarily lead to a substantive outcome; however, it is the first step towards a substantive outcome. In that regard it is important because the initial hurdle has now been removed.

Te Aho suggests that in combination with other elements of the settlement, such as the post-settlement governance entity being recognised as 'having an interest … greater than the public generally when applying the *RMA*', the legal status of the River 'provides the strongest opportunity for more effective participation by Iwi in planning processes of all freshwater settlements to date'.[186]

So legal personhood in isolation does little to improve *iwi* participation; it is only in combination with the rest of the settlement that *iwi* participation is improved, thus providing support for the criticism noted earlier.

LEGAL LIABILITY

Te Awa Tupua not only has the rights and powers of a legal person, but also the liabilities, with Te Pou Tupua being responsible for the liabilities of Te Awa Tupua as a landowner.[187] In that respect Te Pou Tupua can ask for assistance from the Crown to meet its liabilities.[188] Stone noted that '[r]ivers drown people, and flood over and destroy crops',[189] so the possibility of an action being brought against Te Awa Tupua should not be discounted.[190] How the courts would deal with these naturally occurring events is open to speculation, as arguably these are not in the nature of landowner liabilities.

THE LEGAL PERSONALITY MODEL AS A WESTERN LEGAL CONSTRUCT

Finally, one could argue that the legal personality model is an inherently Western one. That is, it might be said that the model is based on Western legal concepts reflected in recognised legal structures, even if Stone's version of the model applies those structures to entities not previously the subject of the model – natural objects. However, the model in the Whanganui River Settlement could be viewed as an attempt to syncretise two different systems, albeit the dominant one still being the non-Māori system. The Whanganui Iwi's struggle for nearly 150 years for recognition of their rights over the Whanganui River has taken place within the non-Māori legal system, the current settlement being the culmination of this lengthy battle. It was a battle that took place within a Western legal construct. However, the Whanganui River Settlement is significant in that it takes what is essentially a Western legal model and endows it with distinctly Māori

186 Te Aho, 'Upholding the Mana of the Whanganui River', above n 171.

187 *Te Awa Tupua Act* s 21(2). There are, however, some exclusions for which the Crown retains liability: s 56, sc 5 cls 1–2.

188 Ibid sch 5 cl 3.

189 Stone, 'Should Trees Have Standing?' (1972), above n 86, 481.

190 See, generally, Erin O'Donnell, 'Legal Rights for Rivers: More Power Less Protection?' on Gabriel Eckstein, *International Water Law Project Blog* (23 April 2018) <www.internationalwaterlaw.org/blog>.

characteristics. In that respect, it goes further than Stone's initial vision in 1972 of a river as a legal entity that was limited to one which aimed to protect purely environmental characteristics.

Other aspects of the settlement of relevance to river management

The preceding discussion has focussed largely on the legal framework that grants legal personality to the Whanganui River. However, there are other parts of the settlement, which may promote the participation of Whanganui Iwi more directly in the management of the Whanganui River.

POST-SETTLEMENT GOVERNANCE ENTITY

The settlement contemplated the establishment of a post-settlement governance entity ('PSGE'), Ngā Tāngata Tiaki o Whanganui, to implement the settlement and to receive, hold and manage any settlement assets and money on trust for the Whanganui Iwi.[191] That entity was established on 4 August 2014.[192] The *Te Awa Tupua Act* recognises that the PSGE as trustee has an interest in Te Awa Tupa that is 'greater than, and separate from, any interest in common with the public generally'.[193] This will give the PSGE legal standing in any litigation and in all relevant statutory processes relating to Te Awa Tupua, an important procedural right.

In relation to the *RMA*, the PSGE will be entitled to lodge submissions in relation to any matter involving the Whanganui River and will be entitled to be heard in relation to any matter involving the Whanganui River.[194] Similar entitlements will apply to other statutory processes.[195] It will also be an *iwi* authority for the purposes of the *RMA*. [196] As an *iwi* authority, a local authority can transfer any of its functions, powers or duties to the PSGE.[197] It also means that it can enter into JMAs with local authorities.[198] However, in practice these provisions of the *RMA* have been little utilised, as local authorities have, to date, been reluctant to make use of them.[199]

Accordingly, whereas Te Pou Tupa will be able to represent the interests of Te Awa Tupua in any proceedings, and indirectly the interests of Whanganui Iwi where they coincide with the interests of Te Awa Tupua, the PSGE will be able to

191 *Ruruku Whakatupua – Te Mana o Te Iwi o Whanganui*, above n 94, ch 10.
192 *Deed of Trust – Ngā Tāngata Tiaki o Whanganui* (4 August 2014) <www.ngatāngatatiaki.co.nz/wp-content/uploads/2015/04/Ng%C4%81-T%C4%81ngata-Tiaki-o-Whanganui-Trust-Deed-Final-Executed-Deed-4-August-2014-1.pdf>.
193 *Te Awa Tupua Act* ss 72(d), 73(1)(d).
194 Ibid s 72(b)–(c).
195 Ibid s 73(1)(b)–(c).
196 Ibid s 72(a)(i).
197 *RMA* s 33.
198 Ibid s 36B.
199 See Chapter 9.

represent the specific interests of the Whanganui Iwi. It is this aspect of the settlement which is of most benefit for the Whanganui Iwi in terms of their direct participation in decisions affecting the management of the Whanganui River.[200]

POTENTIAL FOR CONFLICT

A question arises regarding the recognition of Te Pou Tupua as a public authority and the PSGE as an *iwi* authority. This enables both Te Pou Tupua and the PSGE to seek to have a local authority transfer functions pursuant to section 33 or to enter a JMA pursuant to section 36B of the *RMA*. This creates a potential for conflict if both entities did in fact seek to have a local authority transfer the same functions, or enter a JMA over the same area. This does not appear to have been addressed in the settlement, and it may be an unduly apprehensive view of the practical realities of relationships on the ground. It is a concern which is also tempered by the reluctance of local authorities to transfer powers under s 33 or enter into JMAs under s 36B. Nonetheless, it appears to be theoretically possible, although only time will tell whether it becomes a reality. A similar potential for conflict may also arise with the deeming of Te Awa Tupua as a body corporate for the purposes of applying to be an HPA.

Concluding remarks on the Whanganui River Settlement

The recognition of the Whanganui River as a legal entity, in combination with various elements of the rest of the settlement, is an innovative development in river management by Māori, even with the complexities raised in the discussion above. It is undeniably Māori in its terms; it guarantees standing to access the courts to protect the River's values, those values being unquestionably Māori in orientation. It ensures that the River's values are considered in any decisions which will affect the River or its catchment; it provides for significant Māori representation on the strategy group; it promotes relationships with local authorities; it enhances the ability for heritage protection; it opens the door for the River itself to become a consent authority. The major deficiencies of the legal personality/guardianship model are that it does not fundamentally change the governance of the Whanganui River, and that it only provides for indirect participation by the Whanganui Iwi in the management of the River as the river guardian does not represent Māori, but, instead, represents the River. However, this latter deficiency is one which is largely remedied by the role of the PSGE.

Given the above, the question is whether there is any scope for adapting the Whanganui River personality/guardianship model to the Australian context. Accordingly, our focus now returns to Victoria, where we will also consider the Waikato River model and the relevant *RMA* provisions.

200 Given that the values of Te Awa Tupua and the objectives of the PSGE are likely to be overlapping, their interests will for the most part be complementary, creating an even stronger voice.

Part F

The way forward

11 Reform and recommendations

We now turn to examining options for improving the participation of Indigenous people in Victoria's water management regime. Having already concluded that Aotearoa New Zealand's native title experience is only of modest assistance, it remains to consider whether the *Resource Management Act 1991* (NZ) ('*RMA*') or the Treaty settlement case studies might offer better options for Victoria.

Recognition of Māori interests in the *RMA*

Both the *RMA* and the *Water Act 1989* (Vic) ('*Water Act 1989*') recognise Indigenous interests in water management. The *RMA*, however, is superior to the *Water Act 1989* in the strength of that recognition. The relationship of Māori with their land, waters and *taonga* are matters of national importance which must be recognised and provided for.[1] Further, decision-makers under the *RMA* must have particular regard to *kaitiakitanga* [2] and must take into account the principles of the *Treaty of Waitangi*. [3] If the proposed amendments become law, then Victoria's *Water Act 1989*, on the other hand, will merely require that Aboriginal cultural values and uses are considered in the management of water resources and waterways.[4]

Consultation requirements under the *RMA*

Both the *RMA* and the *Water Act 1989*, if the proposed amendments become law, also require some form of consultation in the preparation of policy statements, strategies and plans. Apart from appointment to some advisory committees, the *Water Act 1989*, however, will be mostly limited to notifying specified Aboriginal parties rather than a specific requirement for consultation.

1 *RMA* s 6(e).
2 Ibid s 7(a). *Kaitiakitanga* is defined in s 2 as 'the exercise of guardianship by the tāngata whenua of an area in accordance with tikanga Māori in relation to natural and physical resources; and includes the ethic of stewardship'.
3 Ibid s 8.
4 Water and Catchment Legislation Amendment Bill 2017 (Vic) cl 4.

Transfers of powers under the *RMA*

There is no equivalent in Victoria of s 33 of the *RMA*. There is currently little political impetus for the *Water Act 1989* or other legislation relating to water management to be amended to allow for the transfer of any powers, duties or functions to Traditional Owners. However, the delegation of powers is already possible under existing legislation, usually with the consent of the Minister.[5] Indeed, delegation has been used in Aotearoa New Zealand as a 'less powerful' alternative to transfers of power under s 33.[6]

It is therefore possible that a particular function could be delegated to specified Aboriginal parties with the consent of the Minister.[7] However, facilitating the appointment of Traditional Owners to governance entities, rather than reliance on delegatory discretions, may have more chance of success, given that the governance entity would retain its non-Indigenous membership.

The *Aboriginal Heritage Act 2006* (Vic) ('*AHA*') would appear to provide for better heritage protection than the heritage protection orders available under the *RMA*, but the *AHA* is still deficient, as was demonstrated in Chapter 7.

Joint management agreements

Australia has had a relatively long history with joint management agreements ('JMAs'), the first being made in the Northern Territory in 1981 over the Gurig National Park.[8] Victoria took until 2009 to adopt this approach, with the amendment of the *Conservation, Forests and Lands Act 1987* (Vic) ('*CFL Act*')[9] to allow for the establishment of Traditional Owner Land Management Boards ('TOLMBs') with the ability to enter into management agreements over various types of public land. The *Traditional Owner Settlement Act 2010* ('*TOSA*') later extended this by amending the *CFL Act* to allow for the preparation of joint management plans ('JMPs'), and to enable the underlying title of jointly managed areas to be transferred to Traditional Owners.[10]

There are, however, some differences worth exploring briefly between the Victorian joint management arrangements, and the joint management arrangements

5 *Catchment and Land Protection Act 1994* (Vic) s 19F(c); *Water Act 1989* ss 33DP(d), 122B(c); 306.

6 Hamish Rennie, Jill Thomson and Tikitu Tutua-Nathan, *Factors Facilitating and Inhibiting Section 33 Transfers to Iwi* (Geography Department, University of Waikato, 2000) 3.

7 See also *Conservation, Forests and Lands Act* 1987 (Vic) s 82H(b) which allows for delegation to TOLMBs.

8 Dermot Smyth, 'Joint Management of National Parks' in Elspeth Young, Jocelyn Davies and Richard Munro Baker (eds), *Working on Country. Contemporary Indigenous Management of Australia's Lands and Coastal Regions* (Oxford University Press, 2001) 75.

9 Pursuant to the *Parks and Crown Land Legislation Amendment (River Red Gums) Act 2009* (Vic).

10 See Chapter 7.

in Aotearoa New Zealand under both the *RMA* and Treaty settlement legislation, and in particular the *Waikato River Settlement Act*.

Under the *RMA*, there is no express statutory limitation on the nature of the land tenure over which a JMA can be made, or the types of decisions which can be the subject of a JMA. In practice, however, the scope of the one JMA reached to date has been narrowly confined.[11] JMAs under the *Waikato River Settlement Act* are mandatory, unlike JMAs under the *RMA*. But although there is scope for extending the content of these JMAs beyond what is statutorily required,[12] this is only by agreement and therefore suffers from the same problems as JMAs under the *RMA*.

Joint management arrangements under the *CFL Act* are statutorily limited to the management of public land (such as parks and reserves), and land transferred in Aboriginal title to traditional owner group entities ('TOGEs') on the basis that the land's status as public land will remain as it was prior to the transfer of title. Thus the main task of TOLMBs under the *CFL Act* is generally to prepare draft JMPs for the management agreement area. The *CFL Act* does potentially allow for other functions to be conferred on the TOLMB by agreement or delegation,[13] but the preparation of the draft JMP is the only statutorily mandated function. Both the JMP and any other function conferred must be consistent with the Act governing the management of the land.[14]

Thus JMAs will always have a relatively narrow scope, either statutorily or in practice.

Another difference is that the JMA provisions of the *RMA* and under the *Waikato River Settlement Act* do not involve co-governance; the JMA parties remain separate and simply work together to manage the relevant resources. In JMAs under the *CFL Act*, a separate co-governance entity is established; the joint management arrangements under the *CFL Act* are therefore a combination of joint management agreements with a form of co-governance. The co-governance arrangements made pursuant to Treaty settlements, such as the Waikato River Authority, are separate from JMAs.

Utilisation of the JMA provisions of the *RMA* has been minimal to date, largely due to their voluntary nature and the reluctance of local authorities to enter into them. Enhanced JMAs (and co-governance) have emerged out of the Treaty settlement process in Aotearoa New Zealand, and the native title claims settlement process or its Victorian equivalent (the *TOSA*) in Victoria. In these circumstances, there is an incentive to negotiate JMAs (and co-governance arrangements) in order to reach a settlement.[15]

11 The JMA between the Taupo District Council and the Tūwharetoa Māori Trust Board only applies to 'notified resource consent applications or private plan changes on or affecting multiply owned Māori land within the Taupo District'. Taupo District Council and Tūwharetoa Māori Trust Board, *Joint Management Agreement* (January 2008) cl 5.0.

12 *Waikato River Settlement Act* ss 42(c), 52.

13 *CFL Act* s 82H(a), (b).

14 Ibid s 82H(c).

15 Note that the initial Yorta Yorta Cooperative Management Agreement arose out of the unsuccessful native title claim and its devastating consequences for the Yorta Yorta

JMAs do have their shortcomings, but an important benefit is that they bring parties together to achieve a common purpose. They function constructively to build relationships and build capacity, thus creating a space for Indigenous voices to be heard.

Waikato River Vision and Strategy

A key feature of the *Waikato River Settlement Act* is the Vision and Strategy, the primary direction-setting document for the Waikato River. The Vision and Strategy resembles a JMP prepared by a TOLMB under the *CFL Act* in that they both set out the management priorities for the relevant area and strategies to achieve them.

However, TOLMBs can only manage public land or Aboriginal title land that was previously public land, so JMPs under the *CFL Act* must be consistent with the relevant legislation applying to the public land in question. Thus the scope of a JMP is not as extensive as the scope of the Vision and Strategy.

An important point evident from the Waikato River Settlement is the stronger legislative status given to the documents relating to the management of the Waikato River, and in particular the Vision and Strategy. The involvement of Māori in the development and approval of these documents is, of course, a key feature. Additionally, the ability of the WRA to appoint Māori commissioners to hear resource consent applications is also an important strengthening of Indigenous participation in decision-making.

We now turn to the Whanganui River Settlement, to see what it has to offer.

Whanganui River Settlement – recognition of a river as a legal entity

Comparisons with the Victorian Environmental Water Holder

It is arguable that Victoria has already edged towards giving legal personality to natural objects, with the establishment in 2010 of the Victorian Environmental Water Holder ('VEWH'), a body corporate with legal capacity to sue and be sued.[16] The role of the VEWH is to manage Victoria's environmental water holdings for the purposes of:

> (a) maintaining the environmental water reserve in accordance with the environmental water reserve objective; and
> (b) improving the environmental values and health of water ecosystems, including their biodiversity, ecological functioning and water quality, and other uses that depend on environmental condition.[17]

people. This was also the impetus behind the amendments to the *CFL Act* and the alternative settlement process in the *TOSA*.

16 *Water Act 1989* s 34DB.

17 Ibid s 33DC.

Although the environmental water reserve has statutory status,[18] it has not, unlike the Whanganui River, been given legal personhood or corporate legal status, that status being reserved for the VEWH. The VEWH does, however, manage and maintain the environmental water reserve. Erin O'Donnell suggests that the VEWH could be 'responsible for, and representative of the "environment" (or at least ecosystem health) for those aquatic ecosystems for which it has the capacity to deliver water'.[19] Although there are environmental water managers in other states,[20] and an environmental water holder at the Commonwealth level,[21] O'Donnell points out that there are some unique features of the VEWH that make it different from these other models,[22] the inference being that it is therefore more likely to be able to represent the environment. The most important of these features are the VEWH's independence from government, and its corporate form, enabling it to sue and be sued. Unlike the other Australian environmental water management models, the Victorian environment Minister has only limited power to give directions to the VEWH.[23] The VEWH also performs its functions in its own right, not on behalf of the Victorian government.[24] And commissioners appointed to the VEWH can only be removed in limited circumstances, essentially for illegal or improper conduct.[25]

The main features that the VEWH has in common with Te Awa Tupua and its guardian Te Pou Tupua are also these two features: its independence from government and the capacity to sue and be sued, arising from its corporate form. Where they differ, however, is in the involvement of Indigenous people, and the inclusion of Indigenous water values. There is no legislative requirement for Indigenous Victorians to be involved in, or become members of, the VEWH. Nor does the objective of the VEWH specifically acknowledge Indigenous values.

One reason for this difference is that the recognition of the Whanganui River as a legal entity occurred as a result of the negotiations for the settlement of the Whanganui Iwi's claim to the river. By contrast, the establishment of the VEWH occurred as a result of environmental concerns.[26] Indigenous interests in environmental water management were not mentioned anywhere during the debates on the Bill to establish the VEWH.[27]

There are no proposals to change the eligibility for membership of the VEWH in the amendments currently before Parliament. The government has, however,

18 Ibid s 4A.
19 Erin O'Donnell, 'Institutional Reform in Environmental Water Management: The New Victorian Environmental Water Holder' (2011) 22 *Journal of Water Law* 73, 84.
20 Ibid 78.
21 *Water Act 2007* (Cth) pt 6.
22 O'Donnell, above n 19, 79.
23 *Water Act 1989* s 33DS.
24 Ibid s 33DC.
25 Ibid s 33DH.
26 Victoria, *Parliamentary Debates*, Legislative Assembly, 26 May 2010, 1921 (Tim Holding, Minister for Water).
27 Water Amendment (Victorian Environmental Water Holder) Bill 2010 (Vic).

appointed for the first time, an Aboriginal Victorian as a commissioner on the VEWH.[28]

Another important difference between the Te Awa Tupua guardianship model and the VEWH is that the VEWH is directly involved in water management, albeit only in relation to environmental water.[29] It relies on partnerships with other water management authorities (such as Catchment Management Authorities) to implement its decisions,[30] but the fundamental responsibility for managing the environmental water holdings remains with the VEWH.[31] This gives the VEWH an advantage over the Te Awa Tupua guardian, which does not have a direct role to play in the management of the Whanganui River.

A final difference to note is that the VEWH performs its functions across the entire state; it is not limited to a particular river or water feature. This means that in making decisions about the management and use of environmental water, it necessarily has to prioritise between river systems.

Accordingly, Victorian Traditional Owners may instead wish to consider the river guardian model in relation to specific rivers, akin to the Whanganui River Settlement. If the Victorian government was prepared to grant legal personality to a specific river, it could incorporate Aboriginal cultural values into the values of the river to be protected. But if Victoria was to go down this path, there are a number of issues that would need to be addressed.

Application to individual rivers

Relationship with Traditional Owner governance structures

Questions might arise about how the river as a legal entity could be accommodated, given the existence of various Indigenous title-holding and land management entities. The Aotearoa New Zealand model does not appear to deal with this issue, despite the Aotearoa New Zealand Treaty settlement landscape having some similar features. For example, PSGEs would appear to be the equivalent of registered native title bodies corporate under the *Native Title Act 1993* (Cth) ('*NTA*') and additionally, in relation to Victoria, under the *TOSA*, TOGEs.

As noted earlier, in Aotearoa New Zealand there is a potential for conflict if both Te Pou Tupua and the PSGE sought to have a local authority transfer some of its functions pursuant to s 33 of the *RMA* in relation to the Whanganui River to either of them, or to reach a JMA with either of them. Although this conflict is unlikely to occur given that s 33 and s 36B have been so little utilised, it remains a technical possibility. In the Victorian context, there is no equivalent of s 33 of the

28 Lisa Neville, 'New Commissioner to Recognise Aboriginal Water Values' (Media Release, 16 November 2011).

29 *Water Act 1989* s 33DD.

30 Ibid s 33DD(d).

31 O'Donnell, above n 19, 81.

RMA, but the conflict could arise in situations where both the TOGE and the river guardian were seeking a management role.

There might also be a conflict between the VEWH and a river guardian, depending on the values of the river that have been identified as requiring protection by the guardian; environmental values do not always equate to Indigenous values.

These conflicts, however, could be avoided provided there is sufficient consultation prior to any recognition of the river as a legal entity.[32]

Nature of settlement agreements

The implementation of native title and Traditional Owner settlements in Victoria is not by way of legislation, but by way of agreements made pursuant to legislation. It is therefore a contractual relationship between the parties that is the basis upon which implementation proceeds. While there is nothing inherently wrong with this kind of arrangement – and, indeed, there are many benefits of agreement making[33] – it is not as powerful as having specific outcomes set out in legislation. The recognition of Te Awa Tupua as a legal entity has occurred though the enactment of legislation. The likelihood of similar settlement legislation being passed in Victoria is more remote, partly because it would require the Victorian government to change the way it implements settlements, partly because it involves a new concept (the river as a legal entity), and partly because it involves scarce water resources which have always been (and continue to be) a politically charged and highly sensitive topic.[34]

Relationship with native title and Traditional Owner rights

As discussed elsewhere,[35] the *NTA* and its Victorian counterpart, the *TOSA*, provide little to native title holders and Traditional Owners by way of management of inland water resources, only rights to take and use water. In Victoria,

32 What sufficient consultation would look like is beyond the scope of this book, but it is more than a right to make submissions or an opportunity to comment.

33 See, eg, Marcia Langton et al (eds), *Honour Among Nations? Treaties and Agreements with Indigenous People* (Melbourne University Press, 2004); Marcia Langton et al (eds), *Settling with Indigenous People: Modern Treaty and Agreement-Making* (Federation Press, 2006).

34 Royce Millar, 'Treasurer Faces Fight Over Pipeline', *The Age* (online), 2 November 2007 <www.theage.com.au/news/national/treasurer-faces-fight-over-pipeline/2007/11/01/1193619061327.html>; Tanya King, 'Solving Water Security: Don't Ignore the Public', *The Conversation* (online) 3 September 2012 <https://theconversation.com/solving-water-security-dont-ignore-the-public-9173>; 'Griffith Protests Over the Murray Darling Basin Plan', *ABC News (Australia)*, 25 January 2011<www.youtube.com/watch?v=dsXLXHQGq-4>.

35 Katie O'Bryan, 'More Aqua Nullius? The *Traditional Owner Settlement Act 2010* (Vic) and the Neglect of Indigenous Rights to Manage Inland Water Resources' (2016) 40(2) *Melbourne University Law Review* 547.

those water rights have been limited in native title determinations to date to domestic and ordinary use,[36] and under the *TOSA*, to personal, domestic and non-commercial communal needs.[37] So giving a river legal personality is likely to have minimal impact on native title and Traditional Owner water rights and vice-versa. Further, as it would occur pursuant to state legislation, if there was any conflict with the Commonwealth *NTA*, pursuant to s 109 of the *Commonwealth Constitution*, the *NTA* would prevail.

Heritage protection

The role of heritage protection authorities ('HPAs') (Registered Aboriginal Parties, 'RAPs', in Victoria) also needs to be considered. As noted earlier, the *Te Awa Tupua Act* provides that Te Awa Tupua is deemed to be a body corporate for the purposes of applying to be an HPA pursuant to the *RMA*. [38] It would appear that the PSGE could also set up a body corporate to apply to become an HPA.[39] Such authorities are able to seek heritage protection orders for the protection of particular places – that is, they are site-specific orders.

The Victorian regime under the *Aboriginal Heritage Act 2006* (Vic) ('*AHA*') is a little different in that a RAP is appointed for a large area rather than a particular site or place. If the RAP is a native title holder or a TOGE with a recognition and settlement agreement under the *TOSA*, the appointed area will be the area in which the group holds native title or has been recognised as the Traditional Owner Group. However, it is very unlikely that a river as a legal entity (represented by a guardian) would satisfy the criteria that the Aboriginal Heritage Council must take into account when making a decision on a RAP application. These include whether the applicant is a native title party, the terms of any native title agreement, and whether the applicant represents the Traditional Owners of the area.[40] Primacy in heritage protection under the *AHA* is clearly to be given to Traditional Owners.

As noted earlier in relation to the Whanganui River Settlement, the guardian of a river with status as a legal entity does not represent Traditional Owners or indeed Aboriginal people; it represents the river. This is likely to preclude it from becoming a RAP. It would also be precluded from becoming a RAP over any area where there are native title holders or Traditional Owners recognised pursuant to a *TOSA* agreement, as there can only be one RAP for such areas.[41] Thus the *AHA*

36 Except for the Wimmera Clans determination, which does not recognise any native title rights in water.
37 *TOSA* s 79 (definition of 'traditional purposes').
38 *Te Awa Tupua Act* s 17(f).
39 Katie O'Bryan, *From Aqua Nullius to Aqua Minimus? The Legal Recognition in Victoria of Indigenous Rights to Participate in the Management of Water Resources – Lessons from Aotearoa-New Zealand* (PhD Thesis, Monash University, 2015) 357, n 244.
40 *AHA* s 151(3).
41 Ibid s 151(2), (2A).

would need to be amended to enable a river as a legal entity to become a RAP, an amendment that is unlikely to be supported by Traditional Owners.

In any event, the *AHA* is largely ineffective as a tool for Traditional Owners to participate in the management of water resources in a holistic way.[42] Therefore, even if the *AHA* were amended to enable a river as a legal entity to be given RAP status over itself (including its catchment), this would not add to the ability of Traditional Owners to manage water resources. It would instead add an extra layer of bureaucracy and potential for confusion and conflict with Traditional Owners.

In summary, giving a river legal status in Victoria would not give it any ability to use the *AHA* to have a role in the management of water resources. And even if the *AHA* were amended to enable this to occur, the *AHA* is limited in what it can achieve, and it would create a potential for conflict with Traditional Owners.

Which of Victoria's rivers might be suitable?

Nonetheless, if the above-mentioned issues are satisfactorily addressed, a further issue would be the identification of suitable river. Rivers that would be particularly appropriate for recognition as a legal entity are arguably those rivers which have less complicated management structures and involve fewer interests. For example, those rivers which fall within the country of a limited number of Traditional Owner groups and Catchment Management Authorities ('CMAs') and which lie entirely within Victoria's external boundary. In that respect, the Mitchell River in East Gippsland might be suitable, as it lies entirely within Victoria and the traditional country of the Gunaikurnai people and involves only one CMA.[43] Similarly, the Wimmera River lies within the traditional country of the Wimmera Clans,[44] and involves only one CMA.[45]

However, given that the Wimmera Clans settled their native title claim in 2005 and the Gunaikurnai people in 2010, and with a number of other claims still waiting to be settled, there may be little incentive for the state to open negotiations with either group in relation to recognising one of these rivers as a legal entity. On the other hand, the native title settlement might provide the basis for opening negotiations. If the state were to agree to do so, it certainly would be an ideal opportunity to test an innovative form of recognition.[46]

The alternative is to find another river similarly suitable for which a settlement under the *NTA* or *TOSA* has not yet been reached. In that regard, having a multiplicity of interests and management responsibilities in a river may not necessarily be an impediment, and may in fact be a sound justification for the creation of a legal entity to represent the entire river. The *Yarra River Protection (Wilip-*

42 O'Bryan, *From Aqua Nullius to Aqua Minimus?*, above n 39, 213 ff.
43 The East Gippsland CMA.
44 The Wotjobaluk, Jaadwa, Jadawadjali, Wergaia and Jupagulk Peoples.
45 The Wimmera CMA.
46 Wimmera Clans have commenced discussions about their native title settlement agreements in light of the *TOSA*, so it is clearly possible: <www.bglc.com.au/native-title-discussion>.

gin Birrarung murron) Act 2017 (Vic), although it does not grant legal personality to the Yarra River (and the Yarra River is not currently the subject of a native title or *TOSA* claim) specifically provides 'for the declaration of the Yarra River and certain public land in its vicinity for the purpose of protecting it as one living and integrated natural entity'[47] with the existence of numerous interests and management responsibilities across a wide range of entities necessitating a coordinated, holistic approach. It should also be noted that the Whanganui River also has a multiplicity of interests and entities with management responsibilities, which was not an impediment to the granting of legal personality to the Whanganui River in the Whanganui River Settlement.

Concluding remarks

The Aotearoa New Zealand case study of the Whanganui River Settlement and the discussion above indicate that the recognition of a river as a legal entity with its associated guardianship structure is not so far removed from existing water governance structures in Victoria, such as the VEWH, as to be entirely beyond contemplation as a means of giving a greater role to Indigenous people in river management. It is therefore not a radical idea, but could instead be viewed as an extension or adaptation of an existing concept. There are a number of potential obstacles and difficulties that would need to be dealt with, including its relationship with Traditional Owner entities and the settlement context in which any such arrangement would be negotiated. An important step towards its realisation would be to have mandated Indigenous representation on the VEWH. As noted earlier, the Victorian government has taken a step towards achieving this goal by appointing an Aboriginal Victorian as a commissioner on the VEWH.[48]

In addition, if the proposed amendments to the *Water Act 1989* and *Catchment and Land Protection Act 1994* become law, this would mean that other stakeholders and water users become exposed to a more prominent role for Indigenous people, and the river as a legal entity concept could then become more acceptable and trialled in legislation for individual rivers or river catchments. In that respect, the *Yarra River Protection Act* provides some evidence that individual rivers and catchments can be the specific subject of legislation.

Other legislative reform

As outlined in Chapter 6, Victoria proposes to amend its water and catchment management legislation to recognise the role of Indigenous people in water management, ensure that Aboriginal cultural values and uses of waterways are considered, provide for representation on advisory committees, and require notification of and consultation with specified Aboriginal parties on various plans and strategies relating to water. But there is still more that could be done,

47 *Yarra River Protection (Wilip-gin Birrarung murron) Act 2017* (Vic) s 1(a).
48 Neville, above n 28.

particularly in relation to representation on governance structures, such as CMAs, water corporations and the VEWH. Given how recent the current proposed amendments are, it may be some time before the Victorian government takes this next step. On the other hand, the proposed amendments, if they become law and are successful in practice, may make it easier to take that next step and include Indigenous representation on governance structures. In the meantime, Aboriginal Victorians are still being treated as a stakeholder, although now an essential one.

Beyond legislation

The above discussion has focussed on legislative reform, legislative recognition being the primary focus of this book. However, as noted at the outset, legislation is just one tool by which Indigenous participation in water management can be strengthened. Enacting or amending legislation can take time, relying as it does on political will. And once enacted, it relies on appropriate resourcing and enforcement for it to work in practice. Accordingly, Indigenous people have used, and continue to use, other avenues to pursue their water management aspirations.

In Australia, for example, there exist various alliances which continue to lobby for their sovereign right to manage their water resources, a number of which have produced their own water policy statements. These alliances include the Murray and Lower Darling Rivers Indigenous Nations (MLDRIN), the Federation of Victorian Traditional Owner Corporations (FVTOC), the North Australian Indigenous Land and Sea Management Alliance (NAILSMA) and the Northern Basin Aboriginal Nations (NBAN).[49] Another important contributor to policy development, but no longer in existence, was the First Peoples Water Engagement Council, established in 2010 to provide advice to the National Water Commission.

Alliances and initiatives like these are intended to maintain pressure on governments to recognise the rights of Indigenous people to participate in water management on their own terms, not simply as another stakeholder to be consulted.

There are also other initiatives, such as the National Cultural Flows Research Project, which describes itself as 'a game-changing research project driven by Indigenous people, for Indigenous people'.[50] With the support of Indigenous organisations (including MLDRIN and NBAN) as well as various government departments and statutory authorities, this project has the potential to have a major impact on the recognition of Indigenous water rights.

Individual Indigenous groups have also taken initiatives to manage water resources on their traditional lands. The Gunditjmara people in Western Victoria, for example, have been involved in the restoration of Lake Condah for many

49 See also, *Mary River Statement* (August 2009), *Garma International Indigenous Water Declaration* (August 2008).

50 National Cultural Flows Research Project, *About the Project* <http://culturalflows.com.au/~culturalflowscom/index.php?option=com_content&view=article&id=16&Itemid=118>.

years, commencing well before their recognition as native title holders of that area.[51]

Overseas in the US, both the *Safe Drinking Water Act*[52] and *Clean Water Act*[53] allow for federally recognised tribes to be 'treated as a state'[54] and given primary responsibility within their respective jurisdictions for various environmental programmes, including enforcement responsibility over public water supplies and the establishment of water quality standards.

In the Canadian context, treaty and comprehensive land claim agreements have been used by First Nations people in Canada to strengthen their governance of water resources through their own laws. For example, the Westbank First Nations Self-Government Agreement gives that nation jurisdiction to manage and regulate water use,[55] and jurisdiction over the supply, treatment, conveyance, storage and distribution of water.[56] Outside of that process, the Nadleh Whut'en and Stellat'en First Nations have produced their own water policy declaration, which sets out their traditional water laws and with which they expect all proponents of developments on their traditional lands to comply.[57]

The initiatives outlined above are just as important to the realisation of Indigenous aspirations for water management as the type of legislative recognition discussed in this book, because they open the door for Indigenous groups to establish and implement water management processes based on their own water management laws.

51 For more detail on the Lake Condah Restoration Project, see <www.gunditjmirring.com/lake-condah-restoration-project>.

52 42 USC § 300j-11 (1986).

53 33 USC § 1377(e) (1987).

54 Environment Protection Agency, *Tribal Assumption of Federal Laws – Treatment as a State (TAS)* <www.epa.gov/tribal/tribal-assumption-federal-laws-treatment-state-tas>. See also Chapter 9, nn 77, 78.

55 Westbank First Nations Self-Government Agreement 2003 cl 136. This Agreement is given the force of law by the *Westbank First Nation Self-Government Act 2004*, SC 2004, c 17.

56 Westbank First Nations Self-Government Agreement 2003 cl 212(b).

57 Gordon Hoekstra, 'Nadleh Whut'en and Stellat'en hereditary leaders proclaim B.C.'s first aboriginal water laws' Vancouver Sun (online) 30 March 2016 <http://vancouversun.com/business/energy/nadleh-whuten-and-stellaten-hereditary-leaders-proclaim-b-c-s-first-aboriginal-water-laws>; Nadleh Whut'en and Stellat'en, *Yinka Dene 'Uza'hné Surface Water Management Policy* (18 March 2016); Nadleh Whut'en and Stellat'en, *Yinka Dene 'Uza'hné Guide to Surface Water Quality Standards* (18 March 2016) <www.carriersekani.ca/news/yinke-dene-uzahne-guide-to-surface-water-quality-standards>.

12 Lessons to be learned

Finally, this chapter articulates the broad conclusions reached from the case study analysis undertaken in this book, and in the process draws out some lessons of wider value. The first section considers the outcomes of the historical overview, the second deals with the analysis of relevant legislation, the third summarises native title, the fourth deals with *Treaty* settlements and the final section enunciates some recommendations for reform. The chapter concludes with some final remarks on the book.

Historical overview

There was negligible participation by Indigenous people in the initial development of Australian water law and policy, both federally and in Victoria. This was reflected in the limited recognition of Indigenous water rights that was a feature of the legislation governing the water management regime in Victoria. Greater participation in recent Commonwealth and Victorian water law reform has resulted in improved recognition of Indigenous water management rights, indicating both legislative and political forces at play.

Māori have had a much stronger voice in the development of Aotearoa New Zealand's water management regime compared with Indigenous people in Victoria's (and Australia's) water management regime. Their voice is also evident in the most recent reforms. This has been attributed to various factors, the *Treaty of Waitangi* and the Waitangi Tribunal being particularly influential in that regard. But other factors, such as the greater power of the Māori vote and Aotearoa New Zealand's political structure, have also had an impact.

Accordingly, it is clear that any assessment of the adequacy of existing water management regimes must be viewed in light of the historic political and social context of the relevant jurisdiction in which the regime evolved. This context will also have an impact on the nature and extent of any potential reforms that might be successfully implemented.

Water, Indigenous rights and environmental legislation in Australia and Aotearoa New Zealand

In Australia, at the Commonwealth level, there are some specific requirements for Indigenous issues to be considered by decision-makers. The *Water Act 2007* (Cth)

('*Water Act 2007*') requires that decision-makers under the Act have regard to Indigenous issues in the preparation of the Murray–Darling Basin Plan and that the Basin Plan provide details about Indigenous uses of Basin water resources. It also requires Water Resource Plans prepared by the Basin states to be consistent with the Basin Plan, thus compelling the states to consider the Indigenous elements of the Basin Plan.

Indigenous representation is also mandated in certain circumstances. The *Water Act 2007* establishes a Basin Community Committee, which must include at least two Indigenous members with expertise in Indigenous matters relevant to Basin water resources.[1] The *Environment Protection and Biodiversity Conservation Act 1999* (Cth) ('*EPBC Act*') establishes an Indigenous Advisory Committee, the entire membership of which is Indigenous. These are important statutory provisions that ensure there is an Indigenous voice in matters relating to those Acts. However, a significant limitation is that these committees are advisory only; they do not make decisions.

Further, given that the respective spheres of operation of the *Water Act 2007* and the *EPBC Act* are limited by virtue of the limitation on Commonwealth powers under the *Constitution*, they are necessarily circumscribed in what they can provide by way of Indigenous participation in water management.

Victoria's primary piece of water legislation, the *Water Act 1989* (Vic) ('*Water Act 1989*'), has been amended in the last decade to make some provision for Indigenous water rights. It is limited, however, to the right to take and use water for traditional purposes by Traditional Owner groups with a natural resource agreement under the *Traditional Owner Settlement Act 2010* (Vic) ('*TOSA*'), rights which are very similar to the rights that Traditional Owners already have under the *Water Act 1989* as ordinary members of the public. Representation on advisory committees established under Victoria's water laws is likely to improve if the proposed amendments become law, but representation on decision-making authorities will remain deficient.

Other avenues, such as land ownership via land rights legislation and the *TOSA*, are generally insufficient on their own to enable participation in water management on that land, other than what is currently available under the *Water Act 1989.*

Indigenous Victorians with recognised Traditional Owner rights over public land under the *TOSA* can negotiate a Land Use Activity Agreement with the state for certain procedural rights which may relate to the management of water resources on that land. However, because these rights are negotiated, the state is unlikely to agree to extend them beyond the scope of existing procedural rights found in the *Water Act 1989* and the *Native Title Act 1993* (Cth) ('*NTA*').

The *Conservation, Forests and Lands Act 1987* (Vic) provides potential for Indigenous participation in water management, by way of joint management agreements over water bodies. But such agreements are limited in their scope and

1 *Water Act 2007* s 202(5)(c).

application, and therefore have the capacity to fragment Indigenous management of water.

The *Aboriginal Heritage Act 2006* (Vic) ('*AHA*'), although having some potential for water management, suffers from various faults rendering it largely ineffective as a tool for Indigenous Victorians to participate meaningfully in water management. A key feature of the *AHA*, Cultural Heritage Management Plans, are mandatory for waterways but only in limited circumstances, such as for high-impact activities, and will only apply to the area of the activity not the length of the waterway. This, once again, potentially fragments Indigenous water management.

By comparison, Aotearoa New Zealand's legislation governing water management, the *Resource Management Act 1991* (NZ) ('*RMA*'), contains much greater recognition of Indigenous interests. These include the consideration of Māori interests in decision-making processes and various voluntary mechanisms which provide opportunities for Māori participation in water management. However, experience to date suggests that the practical effect of many of these mechanisms, particularly those allowing for direct participation, has been minimal. The most effective provisions are those that are mandatory, namely consultation with Māori in the preparation of plans and policy statements. But even these provisions have their limitations.

Both Victoria and Aotearoa New Zealand tend to focus on Indigenous participation in setting strategic direction. Aotearoa New Zealand makes this a requirement under the *RMA*, and in some cases via *Treaty* settlement legislation. Victoria, should the proposed amendments become law, will also soon require Traditional Owner participation in setting strategic direction.

From these case studies, we can see that governments are more likely to legislate to involve Indigenous people in the provision of advice and setting strategic direction, rather than by giving them direct management responsibilities. We can also see that it is important for any legislation enabling Indigenous participation in water management to be coordinated and not result in a fragmentation of management responsibility.

Native title

The recognition of native title in *Mabo* and the *NTA* have not lived up to expectations, particularly in relation to Indigenous participation in the management of water resources. Recognition of water rights under the *NTA* is limited to only those rights capable of being recognised by the common law, which is reflected in the limited scope of water rights recognised in determinations to date. The provisions of the *NTA* relating to procedural rights, as interpreted by the courts, further reduce the capacity of native title holders to participate in decisions regarding the management of water resources. There are also significant barriers, particularly in the more settled parts of Australia, to Indigenous groups obtaining recognition of any native title rights. Therefore, as a mechanism for enhancing participation of Indigenous people in water management, the *NTA* is extremely deficient.

In Aotearoa New Zealand, because the recognition of any native title rights to water is so bound up in the question of ownership (and what is meant by ownership), native title as an avenue for Māori to obtain greater recognition of their rights to participate in the management of water resources is presently quite unsettled.

Native title, therefore, has proven to be an inadequate avenue (in these two countries at least) for Indigenous people to achieve their aspirations regarding water management.

Treaty settlements

However, the same cannot be said in Aotearoa New Zealand of *Treaty* settlements. The first of the two settlements analysed, the Waikato River Settlement, broke new ground in comparison with earlier settlements in relation to the participation of Māori in water management. It has also gone some way to addressing some of the deficiencies in the *RMA* for meeting Māori water management aspirations. Co-governance via equal representation on the Waikato River Authority ('WRA') has given Māori an equal say in certain matters concerning the Waikato River, not least being the WRA's ability to appoint commissioners to conduct hearings on resource consent applications and its role in reviewing the Vision and Strategy, the primary direction-setting document for the Waikato River. The settlement legislation puts the Vision and Strategy (prepared with significant Māori input) at the pinnacle of the strategy and policy hierarchy, therefore strengthening the recognition of Māori interests in the management of the Waikato River. Via the development (and review) of the Vision and Strategy, Māori were, and continue to be, heavily involved in setting the management direction for the Waikato River.

In addition, mandated and enduring joint management agreements require local authorities and the Māori trust established under the settlement to work together to improve Māori engagement in matters relating to the management of the Waikato River.

In the Whanganui River Settlement, the recognition of the Whanganui River as an independent legal entity, represented by a guardian, can be seen as an innovative development for Māori. It is undeniably Māori in its terms, and it guarantees standing in the courts to bring actions to protect the river's values – those values being unquestionably Māori in orientation. It ensures that the river's values are considered in any decisions which will affect the river or its catchment, it provides for significant Māori representation on the group which develops the river strategy, and it promotes relationships with local authorities. It also opens the door for the river to become a heritage protection authority and to have decision-making powers transferred to it in the future. But one of its most important attributes, its independence, is arguably also its major deficiency, insofar as direct Māori participation in water management is concerned. Once the members are appointed to the guardianship entity, they act in the best interests of the river, not of their respective appointees. Nonetheless, in combination with other elements of the settlement, including the role of the Whanganui Iwi's post-settlement governance entity, the benefits of recognising the river as an independent legal entity arguably outweigh the disadvantages.

Reforms

Despite the Aotearoa New Zealand legislation having been developed in a context particular to that country, there are elements of the *RMA* and both the Waikato and Whanganui River settlement statutes that could have wider application in other common law countries, including Australia. One can already see some foundation in the Victorian legislation on which such reforms could build.

For example, the river as a legal entity has some similarities with the Victorian Environmental Water Holder ('VEWH') in that they are both independent entities which act in the best interests of the environment (or an element thereof). It would not be much of a step for Victoria to legislate for mandated Indigenous representation on the VEWH. Environmental water managers exist in other common law countries,[2] so, in the right political context, this concept may gain some traction in those jurisdictions.

Mandatory positions on advisory committees is something that has proven to be politically palatable, no doubt because it involves little change to existing decision-making structures. Mandatory positions on decision-making bodies, such as the VEWH, Water Corporations and Catchment Management Authorities, would also make little change to decision-making structures, and therefore this too is something that could realistically be achieved.

Joint Management Agreements are another viable avenue for meaningful Indigenous participation in water management, provided that the legislation under which they are established enables the holistic management of the water resource.

Innovative approaches to Indigenous water management have come from the settlement of treaty claims whereby the outcome is reflected in legislation. Treaties (currently absent in Australia) and treaty-making processes (currently emerging in Victoria) feature in other countries with a similar common law heritage, including North America. Accordingly, the settlement of treaty claims or the making of new treaties provide an opportunity for Indigenous groups to press for greater water management responsibilities, and in doing so they could consider whether either of the Aotearoa New Zealand approaches may be a way in which they can achieve their water management aspirations.

Whatever path is chosen, elevation above mere participation in public consultation processes is essential to ensure that Indigenous people are not treated as simply another stakeholder providing input into water management decision-making. Their status as First Nations peoples demands it. This could be achieved

2 See, eg, Erin O'Donnell, *Constructing the Aquatic Environment as a Legal Subject: Legal Rights, Market Participation, and the Power of Narrative* (PhD Thesis, University of Melbourne, 2017) ch 6, where she discusses the environmental water managers of western USA.

by making specific provision in water management legislation for Indigenous participation, in the nature of the various legislative reform proposals discussed here.

Final remarks

It must be acknowledged that legislative amendment will not be the panacea for resolving all the issues surrounding Indigenous participation in water management, and there are many examples, some of which were noted earlier in this book, of Indigenous people pursuing non-legislative initiatives to assert their rights to manage water resources on their traditional lands. But the law is still an indispensable tool for effecting change and can complement these other initiatives by providing space for Indigenous voices to be heard. The intention of this book has been to find some practical and achievable ways to make the law more effective for Indigenous people in relation to their water management aspirations. Mandating a voice for Indigenous people in water management legislation recognises that Indigenous people are not simply another stakeholder whose views should be considered. It also redresses an imbalance that has been in existence since the first water laws were enacted. And from this legislative base a stronger Indigenous voice may emerge.

Glossary of Māori terms[1]

hapū	sub tribe, clan
hui	gather, meeting
iwi	tribe
kaitiakitanga	guardianship
kāwanatanga	government
mahinga kai	cultivation, vegetable garden
mana	integrity, charisma, prestige, formal, status, jurisdiction, power, control
mana whakahaera	jurisdiction, administration, control, responsibility
mauri	life principle, special character
rangatiritanga	kingdom, principality, sovereignty, ownership
tāngata whenua	local people, aborigine, native
taonga	property, treasure
tapu	sacred, forbidden, confidential, taboo
wai	water
whakapapa	genealogy, cultural identity

1 In accordance with the New Zealand Style Guide, the meaning of these words is sourced from P M Ryan, *The Raupō Dictionary of Modern Māori* (Penguin, 4th edn, 2012). Where there is also a legislative definition, or where there is some dispute or alternative usage, this is noted in the text.

Bibliography

Articles, books, reports and papers

Anaya, S James, *Indigenous Peoples in International Law* (Oxford University Press, 2nd edn, 2004)

Anaya, S J,'The Emergence of Customary International Law Concerning the Rights of Indigenous Peoples ' (2005) 12 *Law and Anthropology* 127

Anaya, S James, 'Indigenous Peoples' Participatory Rights in Relation to Decisions About Natural Resource Extraction: The More Fundamental Issue of What Rights Indigenous Peoples Have in Lands and Resources' (2005) 22(1) *Arizona Journal of International and Comparative Law* 7

Anaya, James, 'The Emergence of Customary International Law Concerning the Rights of Indigenous Peoples' (2005) 12 *Law and Anthropology* 127

Anaya, S James, *International Human Rights and Indigenous Peoples* (Aspen, 2009)

Anderson, Brigid, Frank Hytten and Clare Land, 'Aboriginal Heritage Bill' (2006) 6(17) *Indigenous Law Bulletin* 16

Application for a Water Conservation Order on Lake Ellesmere – Report of the Committee of the National Water and Soil Conservation Authority (1988)

Arnstein, Sherry R, 'A Ladder of Citizen Participation' (1969) 35(4) *Journal of the American Institute of Planners* 216

Atkinson, Henry, 'Yorta Yorta Nation Aboriginal Corporation's Objection to the Cultural Heritage Bill' (2006) 6(17) *Indigenous Law Bulletin* 18

Attorney-General's Department (Cth), *Reforms to the Native Title Act 1993 (Cth) Options Paper, November 2017*

Attwood, Bain, *The Making of the Aborigines* (Allen & Unwin, 1989)

Attwood, Bain, *Rights for Aborigines* (Allen & Unwin, 2003)

Australian Law Reform Commission, *Connection to Country: Review of the Native Title Act 1993* (Cth), Report No 126 (2015)

Agricultural and Resource Management Council of Australia and New Zealand, *Water Allocations and Entitlements: A National Framework for the Implementation of Property Rights in Water* (Occasional Paper No 1, Task Force on COAG Water Reform, October 1995)

Agricultural and Resource Management Council of Australia and New Zealand and Australian and New Zealand Environment and Conservation Council, *National Principles for the Provision of Water for Ecosystems* (Occasional Paper SWR No 3, Sustainable Water Resources Management Committee Subcommittee on Water Resources, July 1996)

Baker, Jeremy, 'The Waikato-Tainui Settlement Act: A New High Water Mark for Natural Resources Co-management' (2013) 24 *Colorado Journal of International Environmental Law and Policy* 163

Barraclough, Tom, *How Far Can the Te Awa Tupua (Whanganui River) Proposal Be Said to Reflect the Rights of Nature in New Zealand?* (LLB(Hons) Thesis, University of Otago, 2013)

Bartlett, Richard, *Native Title in Australia* (LexisNexis Butterworths, 3rd edn, 2014)

Bartlett, Richard, 'An Obsession with Traditional Laws and Customs Creates Difficulty Establishing Native Title Claims in the South: Yorta Yorta' (2003) 31 *University of Western Australia Law Review* 35

Barwick, Diane, *Rebellion at Coranderrk* (Aboriginal History, 1998)

Behrendt, Jason and Peter Thomson, 'The Recognition and Protection of Indigenous Interests in NSW Rivers' (2004) 3 *Journal of Indigenous Policy* 37

Belich, James, *Making Peoples: A History of New Zealanders – From Polynesian Settlement to the End of the Nineteenth Century* (Allen Lane/Penguin Press, 1996)

Berkes, Fikret, Peter George and Richard J Preston, 'Co-Management: The Evolution in Theory and Practice of the Joint Administration of Living Resources' (1991) 18(2) *Alternatives* 12

Berman, Howard R, 'The International Labour Organization and Indigenous Peoples: Revision of ILO Convention No 107 at the 75th Session of the International Labour Conference, 1988' (1988) 41 *The Review (International Commission of Jurists)* 48

Biggs, Bruce, 'Humpty Dumpty and the Treaty of Waitangi' in I H Kawharu (ed), *Waitangi: Māori and Pakeha Perspectives on the Treaty of Waitangi* (Oxford University Press, 1989) 300

Bjørnlund, Henning, *Water Trade Policies as a Component of Environmentally, Socially and Economically Sustainable Water Use in Rural Southeastern Australia* (PhD Thesis, University of South Australia, 1999)

Black, C L, *The Land is the Source of the Law: A Dialogic Encounter with Indigenous Jurisprudence* (Routledge, 2011)

Blackmore, D J, 'Managing for Scarcity: Water Resources in the Murray Darling Basin' (2000) *CEDA Bulletin* 50

Blackstone, William, *Commentaries on the Laws of England* (Clarendon Press, 2nd edn, 1766)

Boast, Richard P, 'The Waitangi Tribunal: "Conscience of the Nation", or Just Another Court?' (1993) *University of New South Wales Law Journal* 223

Boast, Richard, 'The Treaty of Waitangi and the Law' (1999) *New Zealand Law Journal* 123

Boast, Richard, *Buying the Land, Selling the Land: Governments and Māori Land in the North Island 1865–1921* (Victoria University Press, 2008)

Branson, Catherine, 'The Environmental Protection and Biodiversity Conservation Act 1999 – Some Key Constitutional and Administrative Issues' (1999) 6(1) *Australasian Journal of Natural Resources Law and Policy* 33

Brookfield, F M (Jock), 'The Waitangi Tribunal and the Whanganui River-Bed' (2000) *New Zealand Law Review* 1

Broome, Richard, *Aboriginal Victorians: A History Since 1800* (Allen & Unwin, 2005)

Broome, Richard, *Aboriginal Australians: A History Since 1788* (Allen & Unwin, 4th edn, 2010)

Borrows, John, *Recovering Canada: The Resurgence of Indigenous Law* (University of Toronto Press, 2002)

Borrows, John,(Kegedonce) *Drawing Out Law: A Spirit's Guide* (University of Toronto Press, 2010)

Borrows, John, *Canada's Indigenous Constitution* (University of Toronto Press, 2010)

Buchan, Bruce, '"Aboriginal Welfare" and the Denial of Indigenous Sovereignty' (2002) 20*Arena Journal* 97

Carney, Gerard, *The Constitutional Systems of the Australian States and Territories* (Cambridge University Press, 2006)

Castan, Melissa, 'DRIP Feed: The Slow Reconstruction of Self-Determination for Indigenous Peoples' in Sarah Joseph and Adam McBeth (eds), *Research Handbook on International Human Rights Law* (Edward Elgar, 2010) 492

Castleden, Heather et al, 'Reconciliation and Relationality in Water Res and Management in Canada: Implementing Indigenous Ontologies, Epistemologies, and Methodologies' in Steven Renzetti and Diane P Dupont (eds), *Water Policy and Governance in Canada* (Springer, 2017) 69

Charters, Claire, 'A Self-Determination Approach to Justifying Indigenous Peoples' Participation in International Law and Policy Making' (2010) 17(2) *International Journal on Minority and Group Rights* 215

Chesterman, John and Brian Galligan, *Citizens Without Rights: Aborigines and Australian Citizenship* (Cambridge University Press, 1997)

Christie, Gordon, 'Indigenous Legal Theory: Some Initial Considerations' in Benjamin J Richardson, Shin Iman and Kent McNeil (eds), *Indigenous Peoples and the Law: A Comparative and Critical Perspective* (Hart, 2009) 195

Clark, Elizabeth, 'Section 33 of the Resource Management Act 1991' in Janine Hayward (ed), *Local Government and the Treaty of Waitangi* (Oxford University Press, 2003) 43

Clark, Sandford D and A J Meyers, 'Vesting and Divesting: The Victorian Groundwater Act 1969' (1969–1970) 7 *Melbourne University Law Review* 237

Clark, Sandford and Ian Renard, 'The Riparian Doctrine and Australian Legislation' (1969–1970) 7 *Melbourne University Law Review* 475

Clark, Sandford D, 'The River Murray Question: Part I – Colonial Days' (1971) 8 *Melbourne University Law Review* 11

Clark, Sandford D and Ian A Renard, *The Law of Allocation of Water for Private Use, Volume 1: The Framework of Australian Water Legislation and Private Rights* (Research Project 69/16, Australian Water Resources Council, 1972)

Clark, Sandford D, 'The River Murray Waters Agreement: Peace In Our Time?' (1983) 9 *Adelaide Law Journal* 108

Clark, Sandford D, 'The Murray-Darling Basin: Divided Power, Co-operative Solutions' (2003) 22 *Australian Resources and Energy Law Journal* 67

Coates, Natalie, 'Joint Management in New Zealand: Simply Empty Promises?' (2009) 13 (1) *Journal of South Pacific Law* 32

Connell, Daniel, Stephen Dovers and R Quentin Grafton, 'A Critical Analysis of the National Water Initiative' (2005) 10(1) *Australasian Journal of Natural Resources Law and Policy* 81

Connell, Daniel and R Quentin Grafton, 'Water Reform in the Murray-Darling Basin' (2011) 47(12) *Water Resources Research* 1

Cooke, Robin, 'Introduction' (1990) *New Zealand Universities Law Review* 1

Crawford, James, *Brownlie's Principles of Public International Law* (Oxford University Press, 8th edn, 2012)

Crengle, Diane L, 'Perspectives on Māori Participation Under the Resource Management Act' in Inter-Commission Task Force on Indigenous Peoples, *Indigenous Peoples and Sustainability: Cases and Actions* (International Books, 1997) 344

Cronin, Darryl, 'The Lead Up to the Passage of the Native Title Act' in Toni Bauman and Lydia Glick (eds), *The Limits of Change: Mabo and Native Title 20 Years On* (AIATSIS Research Publications, 2012) 47

Davies, Peter, Susan Lawrence and Jodi Turnbull, 'Harvesting Water on a Victorian Colonial Goldfield' (2011) 29 *Australasian Historical Archaeology* 24

Davies, Peter and Susan Lawrence, 'A Mere Thread of Land: Water Races, Gold Mining and Water Law in Colonial Victoria' (2014) 16 *Journal of Australian Colonial History* 165

Davis, B H, '*New Control Over Natural Water*' (1968) *New Zealand Law Journal* 105

Davis, Megan, 'Indigenous Struggles in Standard-setting: The United Nations Declaration on the Rights of Indigenous Peoples' (2008) 9(2) *Melbourne Journal of International Law* 439

Department of Environment, Land, Water and Planning (Vic), *Water for Victoria: Closing the Loop Feedback Report* (2016)

Department of the Environment, Heritage, Water and the Arts (Cth), *Indigenous Heritage Law Reform – For Discussion* (August 2009)

Department of the Environment, Water, Heritage and the Arts (Cth), *Australia's National Heritage* (2010)

Department of Planning and Community Development, *Discussion Paper: Review of the Aboriginal Heritage Act 2006* (September 2011)

Department of Planning and Community Development, *Summary of Submissions and Consultation – Review of the Aboriginal Heritage Act 2006* (April 2012)

Department of Water Resources Victoria, *Water Law Review* (Discussion Paper, Report No 1, Water Management Report Series, September 1986)

Diesendorf, Mark, 'Reclaiming the Ecologically Sustainable Development Process' (1992) 13(3) *Bogong* 20

Dixon, Jennifer and Aileen Wrathall, 'The Reorganisation of Local Government: Reform or Rhetoric?' (1990) *New Zealand Journal of Geography* 2

Douglas, Heather, 'The Participation of Indigenous Australians in Legal Education 1991–2000' (2001) 24 *University of New South Wales Law Journal* 485

Dovers, Stephen, 'Discrete, Consultative Policy Processes: Lessons from the National Conservation Strategy for Australia and National Strategy for Ecologically Sustainable Development' in Stephen Dovers and Su Wild River (eds), *Managing Australia's Environment* (Federation Press, 2003) 133

D'souza, Lila, 'Native Title Implications for Existing and Future Water Entitlements in Western Australia' (Legal Report, National Native Title Tribunal, 3 May 2002)

Durie, E Taihakurei and Gordon S Orr, 'The Role of the Waitangi Tribunal and the Development of a Bicultural Jurisprudence' (1990) 14 *New Zealand Universities Law Review* 62

Ecologically Sustainable Development Working Groups, *Final Report – Fishing* (November 1991)

Ecologically Sustainable Development Working Groups, *Final Report – Mining* (November 1991)

Ecologically Sustainable Development Working Groups, *Final Report – Tourism* (November 1991)

Ecologically Sustainable Development Working Group Chairs, *Intersectoral Issues Report* (January 1992)

Economou, Nick and Zhareh Ghazarian, *Electoral Redistributions in Theory and Practice* (Report, Victorian Electoral Commission/Monash University, November 2006)

Effective Management Solutions P/L, *Consultations with Indigenous Groups for the Development of the Draft Western Region Sustainable Water Strategy* (28 July 2009)

Erueti, Andrew, 'The Recognition of Indigenous Peoples' Rights to Traditional Lands: The Evaluation of States by International Treaty Bodies' in Claire Charters and Andrew Erueti (eds), *Māori Property Rights and the Foreshore and Seabed: The Last Frontier* (Victoria University Press, 2007) 175

Evatt, Elizabeth, *Review of the Aboriginal and Torres Strait Islander Heritage Protection Act 1984* (22 August 1996)

Evatt, Elizabeth, 'Overview of State and Territory Aboriginal Heritage Legislation' (1998) 4(16) *Indigenous Law Bulletin* 4

Expert Panel, *Report of the Independent Review of the Water Act 2007* (November 2014)

Fisher, D E, *Water Law* (LBC Information Services, 2000)

Fisher, Tim, 'Water Sustainability or Sell-out? The National Water Initiative in Perspective' (2004) 36 *Australian Options* 20

Fitzgerald, Eljon, Brendan Stevenson and Jacob Tapiata, *Māori Electoral Participation – A Report Produced for the Electoral Commission* (Massey University, May 2007)

Foerster, Anita, 'Victoria's New Environmental Water Reserve: What's in a Name?' (2007) 11(2) *Australasian Journal of Natural Resources Law and Policy* 145

Foster, Hamar, Jeremy Webber and Heather Raven (eds), *Let Right Be Done: Aboriginal Title, the Calder Case and the Future of Indigenous Rights* (UBC Press, 2007)

Frame, Alex, 'Property and the Treaty of Waitangi: A Tragedy of the Commodities' in Janet McLean (ed), *Property and the Constitution* (Hart, 1999) 237

Furuseth, Owen and Chris Cocklin, 'An Institutional Framework for Sustainable Resource Management: The New Zealand Model' (1995) 35 *Natural Resources Journal* 243

Gardiner, Greg, *The Aboriginal Heritage Bill 2006* (D-Brief No 2, Parliamentary Library, Parliament of Victoria, 2006)

Gardner, Alex, Richard Bartlett and Janice Gray, *Water Resources Law* (LexisNexis Butterworths, 2nd edn, 2018)

Gibbs, Meredith and April Bennett, 'Māori Claims to Ownership of Freshwater' (2007) *Resource Management Journal* 13

Getches, David H, 'Indigenous Peoples' Rights to Water Under International Norms' (2005) 16 *Colorado Journal of International Environmental Law and Policy* 259

Godden, Lee, 'Water Law Reform in Australia and South Africa: Sustainability, Efficiency and Social Justice' (2005) 17(2) *Journal of Environmental Law* 181

Good, Meg, 'The River as a Legal Person: Evaluating Nature Rights-Based Approaches to Environmental Law' (2013) 1 *National Environmental Law Review* 34

Goodall, Heather, 'Land in Our Own Country: The Aboriginal Land Rights Movement in South-Eastern Australia, 1860–1914' (1990) 14(1) *Aboriginal History* 1

Goot, Murray, 'The Aboriginal Franchise and its Consequences' (2006) 52(4) *Australian Journal of Politics and History* 517

Goulding, Megan, *Northern Region Sustainable Water Strategy: Indigenous Engagement Project* (September 2009)

Gow, L, 'New Zealand Resource Management Act: Implementing A Major Planning Law Reform' in Donna G Craig, Nicholas A Robinson and Koh Kheng-Lian (eds), *Capacity Building for Environmental Law in the Asian and Pacific Region: Approaches and Resources* (Asian Development Bank, 2002) 431

Grant, Jill Elise, 'The Navajo Nation EPA's Experience with "Treatment as a State" and Primacy' (2007) 21(3) *Natural Resources and Environment* 9

Hafner-Burton, Emilie M, 'Sticks and Stones: Naming and Shaming the Human Rights Enforcement Problem' (2008) 62 *International Organization* 689

Hammer, Leonard, 'Indigenous Peoples as a Catalyst for Applying the Human Right to Water' (2003) 10 *International Journal on Minority and Group Rights* 131

Hardcastle, Laura, 'Turbulent Times: Speculations about How the Whanganui River's Position as a Legal Entity Will Be Implemented and How It May Erode the New Zealand Legal Landscape' (February2014) *Māori Law Review* 4

Hare, Bill, *Ecologically Sustainable Development: Assessment of the ESD Working Group Reports* (Australian Conservation Foundation, 1991)

Harris, Edwyna, 'Lobbying for Legislation: An Examination of Water Rights Transition in Colonial Victoria, Australia 1840–1886' (Discussion Paper 12/06, Department of Economics, Monash University, 2006)

Harris, Edwyna, 'Institutional Change and Economic Growth: The Evolution of Water Rights in Victoria, Australia 1850–1886' (2007) 26(2) *Economic Papers* 118

Harris, Edwyna, 'Colonialism and Long-run Growth in Australia: An Examination of Institutional Change in Victoria's Water Sector during the Nineteenth Century' (2008) 48(3) *Australian Economic History Review* 266

Hawke, Allan, *The Australian Environment Act: Report of the Independent Review of the Environment Protection and Biodiversity Conservation Act 1999* (Department of the Environment, Water, Heritage and the Arts, October 2009)

Hayward, Janine, 'Appendix – The Principles of the Treaty of Waitangi' in Volume 2 of Alan Ward, *National Overview of the Waitangi Tribunal Rangahaua Whanui Series* (GP Publications, 1997)

Hemming, Steve et al, 'A New Direction for Water Management? Indigenous Nation Building as a Strategy for River Health' (2017) 22(2) *Ecology and Society* 13

Hickford, Mark, 'Settling Some Very Important Principles of Colonial Law: Three "Forgotten" Cases of the 1840s' (2004) 35 *Victoria University of Wellington Law Review* 1

Hilmer, Frederick G, Mark R Rayner and Geoffery Q Taperell, *National Competition Policy* (Report, Independent Committee of Inquiry, 25 August 1993)

Hitchcock, Robert K, 'International Human Rights, the Environment, and Indigenous Peoples' (1994) 5 *Colorado Journal of International Environmental Law and Policy* 1

Hovell, Devika, 'The Sovereignty Stratagem: Australia's Response to UN Human Rights Bodies' (2003) 28 *Alternative Law Journal* 197

Independent Committee of Review, *Water Law Review: Draft Proposals for a Water Bill* (Report, Department of Water Resources Victoria, January 1989)

Industry Commission, *Water Resources and Waste Water Disposal* (Report No 26, 17 July 1992)

International Alliance of the Indigenous Peoples of the Tropical Forests, 'The Biodiversity Convention: The Concerns of Indigenous Peoples' (1996) 1(4) *Australian Indigenous Law Reporter* 731

Jackson, Sue, *National Indigenous Water Planning Forum: Background Paper on Indigenous Participation in Water Planning and Access to Water. A Report Prepared for the National Water Commission* (CSIRO, 2009)

Johnston, Darlene M, 'A Theory of Crown Trust Towards Aboriginal Peoples' (1986) 18 *Ottawa Law Review* 307

Johnston, Darlene, 'Native Rights as Collective Rights: A Question of Group Self-Preservation' (1989) 2(1) *Canadian Journal of Law and Jurisprudence* 19

Johnston, Darlene, *The Taking of Indian Lands in Canada: Consent or Coercion?* (University of Saskatchewan Native Law Centre, 1989)

Jones, Carwyn, 'A Māori Constitutional Tradition' (2014) 12(1) *New Zealand Journal of Public and International Law* 187

Joseph, Philip A, *Constitutional and Administrative Law in New Zealand* (Brookers, 4th edn, 2014)

Joseph, Robert, 'Reclaiming Legal Space for the First Law of Aotearoa-New Zealand' (2009) 17 *Waikato Law Review* 74

Joseph, Sarah and Joanna Kyriakakis, 'The United Nations and Human Rights' in S Joseph and A McBeth (eds), *Research Handbook on International Human Rights Law* (Edward Elgar, 2010) 1

Kapua, Prue, 'Review of the Role of Māori Under the Resource Management Act 1991' (2007) *Resource Management Theory and Practice* 92

Kelly, Loretta, 'A Personal Reflection on Being a Law Academic' (2005) 6(8) *Indigenous Law Bulletin* 19

Kelly, Nicholas, 'A Bridge? The Troubled History of Inter-State Water Resources and Constitutional Limitations on State Water Use' (2007) 30(3) *University of New South Wales Law Journal* 639

Kelsey, Elizabeth Jane, *Rogernomics and the Treaty of Waitangi* (PhD Thesis, University of Auckland, 1991)

Kennedy, Nathan, 'Viewing the World Through a Wider Lens: Māori and Council Planning Documents' (PUCM Māori Report 6, International Global Change Institute, University of Waikato; and KCSM Consultancy Solutions, 31 July 2008)

Kildea, Paul and George Williams, 'The Constitution and the Management of Water in Australia's Rivers' (2010) 35 *Sydney Law Review* 595

Kildea, Paul and George Williams, 'The Water Act and the Murray-Darling Basin Plan' (2011) 22*Public Law Review* 9

Kingsbury, Benedict, 'Indigenous Peoples in International Law: A Constructivist Approach to the Asian Controversy' (1998) 92 *American Journal of International Law* 414

Kingsbury, Benedict, 'Reconciling Five Competing Conceptual Structures of Indigenous Peoples' Claims in International and Comparative Law' (2001) 34 *New York University Journal of International Law and Politics* 189

Kymlicka, Will, *Multicultural Citizenship: A Liberal Theory of Minority Rights* (Oxford University Press, 1995)

La Nauze, J A, *The Making of the Australian Constitution* (Melbourne University Press, 1972)

Land and Water Forum, *Report of the Land and Water Forum: A Fresh Start for Freshwater* (September 2010)

Land and Water Forum, *Second Report of the Land and Water Forum: Setting Limits for Water Quality and Quantity – Freshwater Policy and Plan-Making through Collaboration* (April 2012)

Land and Water Forum, *Third Report of the Land and Water Forum: Managing Water Quality and Allocating Water* (October 2012)

Langton, Marcia, 'Earth, Wind, Fire and Water: The Social and Spiritual Construction of Water in Aboriginal Societies' in Bruno David, Bryce Barker and Ian J McNiven (eds), *The Social Archaeology of Australian Indigenous Societies* (Aboriginal Studies Press, 2006) 139

Langton, Marcia, Maureen Tehan and Lisa Palmer, 'Introduction' in Marcia Langton et al (eds), *Honour Among Nations? Treaties and Agreements with Indigenous People* (Melbourne University Press, 2004) 1

Langton, Marcia et al (eds), *Honour Among Nations? Treaties and Agreements with Indigenous People* (Melbourne University Press, 2004)

Langton, Marcia et al (eds), *Settling with Indigenous People: Modern Treaty and Agreement-Making* (Federation Press, 2006)

Lavery, Daniel, 'The Participation of Indigenous Australians in Legal Education' (1993) 4 (1) *Legal Education Review* 177

Lebovic, James H and Eric Voeten, 'The Politics of Shame: The Condemnation of Country Human Rights Practices in the UNHCR' (2006) 50 *International Studies Quarterly* 861

Local Government New Zealand, *Co-Management: Case Studies Involving Local Authorities and Māori* (January 2007)

Local Government New Zealand, *Local Authorities and Māori: Case Studies of Local Arrangements* (February 2011)

McAvoy, Tony, 'Water – Fluid Perceptions' (2006) 1(2) *Transforming Cultures eJournal* 97

McGlade, Hannah, 'The Day of the Minstrel Show' (1995) 6(8) *Indigenous Law Bulletin* 16

McHugh, Paul, *The Māori Magna Carta: New Zealand Law and the Treaty of Waitangi* (Oxford University Press, 1991)

Macklem, Patrick, 'Indigenous Recognition in International Law: Theoretical Observations' (2008) 30 *Michigan Journal of International Law* 177

McRae, Heather et al (eds), *Indigenous Legal Issues, Commentary and Materials* (Lawbook, 4th edn, 2009)

Majurey, Paul F et al, *Māori Values Supplement* (Ministry for the Environment, December 2010)

Marshall, Virginia, *Overturning Aqua Nullius: Securing Aboriginal Water Rights* (Aboriginal Studies Press, 2017)

Massola, Aldo, *Coranderrk: A History of the Aboriginal Station* (Lowden Publishers, 1975)

Mazel, Odette, 'The Evolution of Rights: Indigenous Peoples and International Law' (2009) 13(1) *Australian Indigenous Law Review* 140

Mein Smith, Philippa, *A Concise History of New Zealand* (Cambridge University Press, 2nd edn, 2012)

Metcalf, Cherie, 'Indigenous Rights and the Environment: Evolving International Law' (2003) 35 *Ottawa Law Review* 101

Mikaere, Ani, 'Tikanga as the First Law of Aotearoa' (2007) 10 *Yearbook of New Zealand Jurisprudence* 24

Miller, Jane, 'A Guide to Government Acts, Reports and Regulations Relating to Indigenous Victorians' (2010) 85 *La Trobe Journal* 172

Miller, Robert J et al, *Discovering Indigenous Lands: The Doctrine of Discovery in the English Colonies* (Oxford University Press, 2010)

Ministry for the Environment (NZ), *Freshwater for a Sustainable Future: Issues and Options – A Public Discussion Paper on the Management of New Zealand's Freshwater Resources* (December 2004)

Ministry for the Environment (NZ), *Testing the Water: Report on the Sustainable Water Program of Action Written Submissions* (July 2005)

Ministry for the Environment (NZ), *Wai Ora: Report of the Sustainable Water Programme of Action Consultation Hui* (July 2005)

Ministry for the Environment (NZ), *Report of the Sustainable Water Program of Action Public Meetings* (July 2005)

Ministry for the Environment (NZ), *Report of the Sustainable Water Program of Action Local Government Workshops* (July 2005)

Ministry for the Environment (NZ), *Proposed Amendments to the National Policy Statement for Freshwater Management 2011: A Discussion Document* (November 2013)

Ministry for the Environment (NZ), *Report and Recommendations on the Proposed Amendments to the National Policy Statement for Freshwater Management and Public Submissions* (July 2014)

Ministry for the Environment (NZ), *Submissions Report and Recommendations on Proposed Amendments to the National Policy Statement for Freshwater Management 2014* (2017)

Monture, Patricia A, 'Ka-Nin-Geh-Heh-Gah-E-Sa-Nonh-Yah-Gah' (1988) 2 *Canadian Journal of Women and Law* 159

Monture, Patricia A, 'Now that the Door is Open: First Nations and the Law School Experience' (1990) 15 *Queen's Law Journal* 179

Monture, Patricia, 'Reflecting on Flint Woman' in Richard F Devlin (ed), *Canadian Perspectives on Legal Theory* (Emond Montgomery, 1991) 351

Monture-Angus, Patricia, *Thunder in My Soul: A Mohawk Woman Speaks* (Fernwood, 1995)

Monture-Angus, Patricia, *Journeying Forward: Dreaming Aboriginal Peoples' Independence* (Pluto Press, 2000)

Morris, James Douglas Kahotea, *Affording New Zealand Rivers Legal Personality: A New Vehicle for Achieving Māori Aspirations in Co-management?* (LLM Thesis, University of Otago, 2009)

Morris, James D K and Jacinta Ruru, 'Giving Voice to Rivers: Legal Personality as a Vehicle for Recognising Indigenous Peoples' Relationships to Water?' (2010) 14 *Australian Indigenous Law Reporter* 49

Mulligan, Helen and John J Pigram, *Water Administration in Australia: Agenda for Change* (Occasional Paper No 4, University of New England, 1989)

Murray–Darling Basin Authority, *A Yarn on the River: Getting Aboriginal Voices into the Basin Plan* (2011)

Murray–Darling Basin Authority, *Handbook for Practitioners: Water Resource Plan Requirements* (October 2013)

Murray–Darling Basin Commission, *Review of the Operation of the Cap* (Overview Report, August 2000)

National Water Commission, *Australia's Water Blueprint: National Reform Assessment 2014*

Neal, Eric, *Report of the Working Group on Water Resource Policy to the Council of Australian Governments* (1994)

Nettheim, Garth, Gary D Meyers and Donna Craig, *Indigenous Peoples and Governance Structures: A Comparative Analysis of Land and Resource Management Rights* (Aboriginal Studies Press, 2002)

New Zealand Law Commission, *Māori Custom and Values in New Zealand Law* (Study Paper 9, 2001)

Nolan, Derek (ed), *Environmental and Resource Management Law* (LexisNexis, 5th edn, 2015)

O'Bryan, Katie, 'Issues in Natural Resource Management – Inland Water Resources – Implications of Native Title and the Future of Indigenous Control and Management of Inland Waters' (2007) 14(2) *Murdoch University E-law Journal* 280

O'Bryan, Katie, 'The National Water Initiative and Victoria's Legislative Implementation of Indigenous Water Rights' (2012) 7(29) *Indigenous Law Bulletin* 24

O'Bryan, Katie, *From Aqua Nullius to Aqua Minimus? The Legal Recognition in Victoria of Indigenous Rights to Participate in the Management of Water Resources – Lessons from Aotearoa-New Zealand* (PhD Thesis, Monash University, 2015)

O'Bryan, Katie, 'Victoria's Traditional Owner Settlement Act and Indigenous Management of Water Resources: An Improvement on the Native Title Act or More of the Same?' (2016) 8(26) *Indigenous Law Bulletin* 21

O'Bryan, Katie, 'More Aqua Nullius? The Traditional Owner Settlement Act 2010 (Vic) and the Neglect of Indigenous Rights to Manage Inland Water Resources' (2016) 40(2) *Melbourne University Law Review* 547

O'Donnell, Erin, 'Institutional Reform in Environmental Water Management: The New Victorian Environmental Water Holder' (2011) 22*Journal of Water Law* 73

O'Donnell, Erin and Julia Talbot-Jones, 'Legal Rights for Rivers: What Does This Actually Mean?' (2017) *Australian Environment Review* 159

O'Donnell, Erin, *Constructing the Aquatic Environment as a Legal Subject: Legal Rights, Market Participation, and the Power of Narrative* (PhD Thesis, University of Melbourne, 2017)

O'Donnell, Michael, *NAILSMA – TRaCK Project 6.2: Indigenous Rights in Water in Northern Australia* (Charles Darwin University, 2011)

Orange, Claudia, *The Treaty of Waitangi* (Allen & Unwin, 1987)

Palmer, Geoffrey and Matthew Palmer, *Bridled Power: New Zealand's Constitution and Government* (Oxford University Press, 4th edn, 2004)

Palmer, Kenneth A, *Local Government Law in New Zealand* (Law Book, 2nd edn, 1993)

Palmer, Kenneth A, *Local Authorities Law in New Zealand* (Brookers, 2012)

Palmer, Matthew, 'The Treaty of Waitangi in Legislation' (2001) *New Zealand Law Journal* 207

Palmer, Matthew S R, *The Treaty of Waitangi in New Zealand's Law and Constitution* (Victoria University Press, 2008)

Parker, M L, T J Hundloe and D R Chapman, 'Water Resources and Waste Water Disposal' (Report No 26, Industry Commission, 17 July 1992)

Parliamentary Commissioner for the Environment, *Hydroelectricity or Wild Rivers? Climate Change Versus Natural Heritage* (May 2012)

Parrish, Austin L, 'Changing Territoriality, Fading Sovereignty, and the Development of Indigenous Rights' (2007) 31*American Indian Law Review* 291

Pigram, John J, *Australia's Water Resources: From Use to Management* (CSIRO Publishing, revised edn, 2007)

Powell, J M, *Environmental Management in Australia, 1788–1914: Guardians, Improvers and Profit: An Introductory Survey* (Oxford University Press, 1976)

Powell, J M, *Watering the Garden State: Water, Land and Community in Victoria 1834–1988* (Allen & Unwin, 1989)

Pyle, Eric et al, 'Establishing Water Management in Law: The New Zealand Experience' (2001) 37(4) *Journal of the American Water Resources Association* 783

Rennie, Hamish, Jill Thomson and Tikitu Tutua-Nathan, *Factors Facilitating and Inhibiting Section 33 Transfers to Iwi* (Geography Department, University of Waikato, 2000)

Report by a Special Tribunal Appointed by the Minister for the Environment to Consider an Application for a Water Conservation Order for the Oreti River (November 2007)

Reynolds, Henry, *The Other Side of the Frontier: Aboriginal Resistance to the European Invasion of Australia* (University of NSW Press, first published 1981, 2006 edn)

Reynolds, Henry, *Frontier: Aborigines, Settlers and Land* (Allen & Unwin, first published 1987, 1996 edn)

Reynolds, Henry, *With the White People* (Penguin, 1990)

Rodgers-Falk, Phillip, *Growing the Number of Aboriginal and Torres Strait Islander Law Graduates: Barriers to the Profession* (September 2011) <https://docs.education.gov.au/system/files/doc/other/rodgers-falk_2011.pdf>

Ross, R M, 'Te Tiri o Waitangi' (1972) 6 *New Zealand Journal of History* 133

Ruru, Jacinta, 'The Māori Encounter with Aotearoa New Zealand's Legal System' in Benjamin J Richardson, Shin Imai and Kent McNeil (eds), *Indigenous Peoples and the Law: Comparative and Critical Perspectives* (Hart, 2009) 111

Ruru, Jacinta, 'Undefined and Unresolved: Exploring Indigenous Rights in Aotearoa New Zealand's Freshwater Legal Regime' (2009) 20 *Journal of Water Law* 236

Ruru, Jacinta, 'The Legal Voice of Māori in Freshwater Governance – A Literature Review' (Landcare Research New Zealand, October 2009)

Ruru, Jacinta, 'Claiming Title in the Foreshore and Seabed' in Louis A Knafla and Haijo Westra (eds), *Aboriginal Title and Indigenous Peoples: Canada, Australia and New Zealand* (UBC Press, 2010) 185

Ruru, Jacinta, 'Indigenous Restitution in Settling Water Claims: The Developing Cultural and Commercial Redress Opportunities in Aotearoa, New Zealand' (2013) 22 *Pacific Rim Law and Policy Journal* 311

Sambo, Dalee, 'Indigenous Peoples and International Standard Setting Processes: Are State Governments Listening?' (1993) 3(1) *Transnational Law and Contemporary Problems* 13

SAMLIV Project Team, *Strategy for Aboriginal Managed Lands in Victoria* (SAMLIV Steering Committee, December 2003)

Senate Environment, Communications, Information Technology and the Arts References Committee, Parliament of Australia, *Commonwealth Environment Powers* (1999)

Sharpe, Andrew, *Justice and the Māori: The Philosophy and Practice of Māori Claims in New Zealand since the 1970s* (Oxford University Press, 3rd edn, 1997)

Sheppard, David et al, *Report and Recommendations of the Board of Enquiry into the Proposed National Policy Statement for Freshwater Management* (2010)

Smelcer, John D, 'Using International Law More Effectively to Secure and Advance Indigenous Peoples' Rights: Towards Enforcement in U.S. and Australian Domestic Courts' (2006) 15 *Pacific Rim Law and Policy Journal* 301

Smith, David Ingle, *Water in Australia: Resources and Management* (Oxford University Press, 1998)

Smith, Dingle, 'Water Resources Management' in Stephen Dovers and Su Wild River (eds), *Managing Australia's Environment* (Federation Press, 2003) 53

Smith, Laurajane, 'A History of Aboriginal Heritage Legislation in South-Eastern Australia' (2000) 50(1) *Australian Archaeology* 109

Smyth, Dermot, 'Joint Management of National Parks' in Elspeth Young, Jocelyn Davies and Richard Munro Baker (eds), *Working on Country. Contemporary Indigenous Management of Australia's Lands and Coastal Regions* (Oxford University Press, 2001) 75

'Special Focus Edition: Racism in Legal Education' (2005) 6(8) *Indigenous Law Bulletin*

Spiller, Peter, Jeremy Finn and Richard Boast, *A New Zealand Legal History* (Brookers, 2nd edn, 2001)

State Rivers and Water Supply Commission, *Underground Water Resources of Victoria: Report of the State Development Committee – Analysis of Recommendations* (Melbourne, 1962)

Steering Committee for the Development of a Native Title Settlement Framework, *Report of the Steering Committee for the Development of a Native Title Settlement Framework* (December 2008)

Stephenson, Janet, 'Recognising Rangatiratanga in Resource Management for Māori Land: A Need for a New Set of Arrangements' (2001) 5 *New Zealand Journal of Environmental Law* 159

Stoeckel, Kate, *Australian Water Law* (Thomson Reuters, 2012)

Stone, Christopher, 'Should Trees Have Standing? Toward Legal Rights for Natural Objects' (1972) 45 *Southern California Law Review* 450

Strelein, Lisa, *Compromised Jurisprudence. Native Title Cases Since Mabo* (Aboriginal Studies Press, 2006)

Sutton, J D, 'The Treaty of Waitangi Today' (1981) *Victoria University of Wellington Law Review* 17

Tan, Poh-Ling, 'Irrigators Come First: Conversion of Existing Allocations to Bulk Entitlements in the Goulburn and Murray Catchments, Victoria' (2001) 18(2) *Environmental and Planning Law Journal* 154

Tan, Poh-Ling 'The Changing Concepts of Property in Surface Water Resources in Australia' (2002) 13(4) *Journal of Water Law* 269

Tan, Poh-Ling, 'Legal Issues Relating to Water Use' in Catherine Mobbes and Ken Moore (eds), *Property: Rights and Responsibilities – Current Australian Thinking* (Land and Water Australia, 2002) 13

Tan, Poh-Ling, 'Legislating for Adequate Public Participation in Allocating Water in Australia' (2006) 31(1) *Water International* 12

Tan, Poh-Ling et al, *Collaborative Water Planning: Context and Practice – Literature Review, Volume 1* (TRaCK Report, September 2008)

Te Aho, Linda, 'Contemporary Issues in Māori Law and Society' (2005) 13 *Waikato Law Review* 145

Te Aho, Linda, 'Contemporary Issues in Māori Law and Society: The Tangled Web of Treaty Settlements – Emissions Trading, Central North Island Forests, and the Waikato River' (2008) 16 *Waikato Law Review* 229

Te Aho, Linda, 'Indigenous Challenges to Enhance Freshwater Governance and Management in Aotearoa New Zealand: The Waikato River Settlement' (2009) 20 *Journal of Water Law* 285

Te Aho, Linda, 'Indigenous Aspirations and Ecological Integrity: Restoring and Protecting the Health and Wellbeing of an Ancestral River for Future Generations in Aotearoa New Zealand' in Laura Westra, Klaus Bosselman and Colin Suskolne (eds), *Globalisation and Ecological Integrity in Science and International Law* (Cambridge Scholars, 2011) 346

Te Aho, Linda, 'Nga Whakataunga Waimaori – Freshwater Settlements' in Nicola Wheen and Janine Hayward (eds), *Treaty of Waitangi Settlements* (Bridget Williams Books, 2012) 102

Te Aho, Linda, 'Ruruku Whakatupua Te Mana o te Awa Tupua – Upholding the Mana of the Whanganui River' (2014) *Māori Law Review*

Tehan, Maureen, 'A Hope Disillusioned, an Opportunity Lost? Reflections on Common Law Native Title and Ten Years of the Native Title Act' (2003) 27 *Melbourne University Law Review* 523

Temm, Paul, *The Waitangi Tribunal: The Conscience of the Nation* (Random Century, 1990)

Thornberry, Patrick, *Indigenous Peoples and Human Rights* (Manchester University Press, 2005)

Toki, Valmaine, 'Tikanga Māori – A Constitutional Right? A Case Study' (2014) 40(1) *Commonwealth Law Bulletin* 32

Tomas, Nin, 'Māori Concepts of Rangatiratanga, Kaitiakitanga, the Environment, and Property Rights' in David Grinlinton and Prue Taylor (eds), *Property Rights and Sustainability: The Evolution of Property Rights to Meet Ecological Challenges* (Martinus Nijhoff, 2011) 219

Torres, Raidza, 'The Rights of Indigenous Populations: The Emerging International Norm' (1991) 16 *Yale Journal of International Law* 127

Toussaint, Sandy, Patrick Sullivan and Sarah Yu, 'Water Ways in Aboriginal Australia: An Interconnected Analysis' (2005) 15(1) *Anthropological Forum* 61
Tuhiwai Smith, Linda, *Decolonising Methodologies: Research and Indigenous Peoples* (Zed Books, 2nd edn, 2012)
Turner, Dale, *This Is Not a Peace Pipe: Towards a Critical Indigenous Philosophy* (University of Toronto Press, 2006)
Turpel, Aki-Kwe/Mary Ellen, 'Aboriginal Peoples and the Canadian Charter of Rights and Freedoms: Contradictions and Challenges' (1989) 10 *Canadian Woman Studies* 149
Turpel, Mary Ellen, 'Indigenous Peoples' Rights of Political Participation and Self-Determination: Recent International Legal Developments and the Continuing Struggle for Recognition' (1992) 25 *Cornell International Law Journal* 579
Victorian Catchment Management Council, *A VCMC Position/Perspective Paper on Indigenous Engagement in Victoria* (2006)
Victorian Government, *Securing Our Water Future: Green Paper for Discussion* (Department of Sustainability and Environment, August 2003)
Vince, Jenny, 'Māori Consultation Under the Resource Management Act and the 2005 Amendments' (2006) 10 *New Zealand Journal of Environmental Law* 295
Waitangi Tribunal, *Report of the Waitangi Tribunal on the Kaituna River Claim* (1984)
Waitangi Tribunal, *Muriwhenua Report* (1997)
Waitangi Tribunal, *Whanganui River Report* (1999)
Waitangi Tribunal, *Ko Aotearoa Tēnei: A Report into Claims Concerning New Zealand Law and Policy Affecting Māori Culture and Identity – Te Taumata Tuarua* (Report of Wai 262 Tribunal, 2011)
Waitangi Tribunal, *Interim Report on the National Freshwater and Geothermal Resources Claim* (2012)
Waitangi Tribunal, *Stage 1 Report on the National Freshwater and Geothermal Resources Claim* (2012)
Waitangi Tribunal, *Waitangi Tribunal Bibliography: Tribunal Reports, Publications and Research Reports Presented in Evidence 1975–2012 – Part 1: Waitangi Tribunal Reports, Statements and Publications* (2013)
Water for Victoria: Discussion Paper (Department of Environment, Land, Water and Planning, 2016)
Watson, Irene, 'Some Reflections on Teaching Law: Whose Law, Yours or Mine?' (2005) 6 (8) *Indigenous Law Bulletin* 23
Watson, Irene, *Aboriginal Peoples, Colonialism and International Law* (Routledge, 2015)
Watson, Nicole, 'Indigenous People in Legal Education: Staring into a Mirror without Reflection' (1995) 6(8) *Indigenous Law Bulletin* 4
Weir, Jessica, *The Gunditjmara Land Justice Story* (Native Title Research Unit, AIATSIS, 2009)
Wevers, Samuel George, 'Recognising Rangatiratanga Through Co-Management: The Waikato River Settlement' (2013) 4 *New Zealand Law Review* 689
Wheen, Nicola, 'A Natural Flow: The History of Water Law in New Zealand' (1997) 9(1) *Otago Law Review* 71
Wheen, Nicola, 'The Resource Management Act 1991: A Greener Law for Water?' (1997) 1 *New Zealand Journal of Environmental Law* 165
Wheen, Nicola, 'An Updated History of New Zealand Environmental Law' in Eric Pawson and Tom Brooking (eds), *Making a New Land: Environmental Histories of New Zealand* (Otago University Press, revised edn, 2013) 277
White, Ben, *Inland Waterways: Lakes* (National Theme Q, Waitangi Tribunal Rangahua Whanui Series, March 1998)

Williams, David V, 'Wi Parata is Dead, Long Live Wi Parata' in Claire Charters and Andrew Erueti (eds), *Māori Property Rights and the Foreshore and Seabed: The Last Frontier* (Victoria University Press, 2007) 31

Williams, David V, 'Ko Aotearoa Tēnei: Law and Policy Affecting Māori Culture and Identity' (2013) 20 *International Journal of Cultural Property* 311

Williams, Joe, 'Treaty Making in New Zealand' in Marcia Langton et al (eds), *Honour Among Nations? Treaties and Agreements with Indigenous People* (Melbourne University Press, 2004) 163

Williams, Jr, Robert A, *The American Indian in Western Legal Thought: The Discourses of Conquest* (Oxford University Press, 1990)

Williams, Jr, Robert A, 'Vampires Anonymous and Critical Race Practice' (1997) 95 *Michigan Law Review* 741

Willington, David, *Technical Report No 1 – Indigenous Engagement Summary* (4 May 2011)

Wright, Raymond, *A People's Counsel – A History of the Parliament of Victoria: 1856–1990* (Oxford University Press, 1992)

Young, Simon, *The Trouble with Tradition: Native Title and Cultural Change* (Federation Press, 2008)

Yu, Sarah, *Ngapa Kunangkul: Living Water. Report on the Aboriginal Cultural Values of Groundwater in the La Grange Sub-basin* (University of Western Australia, 2nd edn, May 2000)

Treaties

Charter of the United Nations

Convention on Biological Diversity, opened for signature 5 June 1992, 1760 UNTS 79 (entered into force 29 December 1993)

Convention on the Control of Transboundary Movements of Hazardous Wastes and Their Disposal, opened for signature 20 March 1989, 1763 UNTS 57 (entered into force 5 May 1992)

International Covenant on Civil and Political Rights, opened for signature 16 December 1966, 999 UNTS 171 (entered into force 23 March 1976)

International Covenant on Economic, Social and Cultural Rights, opened for signature 16 December 1966, 993 UNTS 3 (entered into force 3 November 1976)

ILO Convention No 107, *Concerning the Protection and Integration of Indigenous and Other Tribal and Semi-Tribal Populations in Independent Countries*, opened for signature 26 June 1957 (entered into force 2 June 1959)

ILO Convention No 169, *Concerning Indigenous and Tribal Peoples in Independent Countries*, opened for signature 27 June 1989 (entered into force 5 September 1991)

Treaty of Waitangi

United Nations materials

Anaya, James, *Promotion and Protection of All Human Rights, Civil, Political, Economic, Social and Cultural Rights, Including the Right to Development: Report of the Special Rapporteur on the Situation of Human Rights and Fundamental Freedoms of Indigenous People*, UN Doc A/HRC/12/34 (15 July 2009)

Committee on the Elimination of Racial Discrimination, General Recommendation No 23: Indigenous Peoples, 51st sess, UN Doc A/52/18, annex V at 122 (18 August 1997)

Conference Report of the World Commission on Environment and Development, *Our Common Future*, UN Doc A/42/427 (4 August 1987)
ECOSOC Res 1982/34, UN ESCOR, 1st sess, 28th plen mtg, Supp No 1, UN Doc E/1982/82 (7 May 1982)
Expert Mechanism on the Rights of Indigenous Peoples, *Progress Report on the Study on Indigenous Peoples and the Right to Participate in Decision-Making* (2010) UN Doc A/HRC/15/35 (23 August 2010)
Expert Mechanism on the Rights of Indigenous Peoples, *Final Report of the Study on Indigenous Peoples and the Right to Participate in Decision-Making*, UN Doc A/HRC/18/42 (17 August 2011)
HRC Sub-Commission Res 1985/22, 38th sess, 37th mtg, UN Doc E/CN.4/1985/5; E/CN.4/Sub.2/1985/57 (4 November 1985)
Human Rights Committee, Views: Communication No 1457/2006, 95th sess, UN Doc CCPR/C/95/D/1457/2006 (24 April 2009) ('Poma Poma v Peru')
Human Rights Committee, 50th sess, *General Comment No 23: The Rights of Minorities* (Article 27), UN Doc CCPR/C/21/Rev.1/Add.5 (8 April 1994)
Intergovernmental Negotiating Committee for a Convention on Biological Diversity, *Report of the Intergovernmental Negotiating Committee for a Convention on Biological Diversity on the Work of its Fourth Negotiating Session/Second Session*, UN Doc UNEP/Bio.Div/N4-INC.2/5 (2 October 1991)
Intergovernmental Negotiating Committee for a Convention on Biological Diversity, *Report of the Intergovernmental Negotiating Committee for a Convention on Biological Diversity on the Work of its Third Session/Fifth Negotiating Session*, UN Doc UNEP/Bio.Div/N5-Inc.3/4 (4 December 1991)
Intergovernmental Negotiating Committee for a Convention on Biological Diversity, *Report of the Intergovernmental Negotiating Committee for a Convention on Biological Diversity on the Work of its Sixth Negotiation Session/Fourth Session of INC*, UN Doc UNEP/Bio.Div/N6-INC.4/4 (18 February 1992)
International Labour Conference, *Provisional Record 32, Seventy-sixth Session, Geneva 1989, Thirty-sixth Sitting, Tuesday, 27 June*
International Labour Organization, *Indigenous and Tribal Peoples' Rights in Practice: A Guide to ILO Convention No 169* (2009)
International Labour Organization, *Understanding the Indigenous and Tribal Peoples Convention, 1989 (No. 169) Handbook for ILO Tripartite Constituents* (2013)
International Labour Organization, *Ratifications of C169 – Indigenous and Tribal Peoples Convention, 1989 (No. 169)* <www.ilo.org/dyn/normlex/en/f?p=NORMLEXPUB:11300:0::NO:11300:P11300_INSTRUMENT_ID:312314:NO>
Martínez Cobo, José R, *Study of the Problem of Discrimination Against Indigenous Populations*, UN Docs E/CN.4/Sub.2/1986/7 and E/CN4/Sub.2/1986/7/Add.4 (1986)
Report of the Committee on the Elimination of Racial Discrimination, GAOR 54th sess, 1331st mtg, Supp 18, UN Doc A/54/18 (29 September 1999)
Report of the United Nations Conference on Environment and Development, Volume I, Resolutions Adopted by the Conference, UN Doc A/CONF.151/Rev.1 (Vol 1) (adopted 14 June 1992)
Report of the World Commission on Environment and Development, UN GA, 42nd sess, UN Doc A/42/427 (4 August 1987)
United Nations Declaration on the Rights of Indigenous Peoples, GA Res 61/295, UN GAOR, 61st sess, 107th plen mtg, Supp No 49, UN Doc A/RES/61/295 (13 September 2007)

Working Group on Indigenous Populations, *Report of the Working Group on Indigenous Populations on its First Session*, UN Doc E/CN.4/Sub.2/1982/33 (25 August 1982)

Other

Aboriginal Affairs and Northern Development Canada, 'Canada's Statement of Support on the United Nations Declaration on the Rights of Indigenous Peoples, November 12, 2010' (30 July 2012) <www.aadnc-aandc.gc.ca/eng/1309374239861/1309374546142>

Aboriginal and Torres Strait Islander Commission, *Annual Report 2003–2004* (22 October 2004)

Aboriginal Heritage Council, Decision of the Victorian Aboriginal Heritage Council in Relation to an Application by Wurundjeri Tribe Land and Compensation Cultural Heritage Council Inc to be a Registered Aboriginal Party (22 August 2008) <www.vic.gov.au/aboriginalvictoria/heritage/registered-aboriginal-parties/wurundjeri-land-and-compensation-cultural-heritage-council-aboriginal-corporation.html>

Aboriginal Victoria, *Registered Aboriginal Parties: Guidelines for Applicants* (nd), <www.vic.gov.au/aboriginalvictoria/heritage/heritage-tools-and-publications/guides-forms-and-practice-notes-for-aboriginal-heritage-management.html#J>

Aboriginal Victoria, *Aboriginal Heritage Act 2006 Practice Note: Significant Ground Disturbance*, <www.vic.gov.au/aboriginalvictoria/heritage/heritage-tools-and-publications/guides-forms-and-practice-notes-for-aboriginal-heritage-management.html>

Annexure A: Draft Rules of Gippsland and East Gippsland Aboriginal Co-Operative Limited. Registered as a non-trading Co-operative with shares under the Co-operatives Act 1996 (Vict.)

Archer, William Henry, Registrar General, *Census of Victoria, 1861. Population Tables – Part 1 – Inhabitants and Houses: Showing the Numbers and Distribution of the People* (Melbourne, 25 April 1862)

Australian Bureau of Statistics, *4705.0 Population Distribution, Aboriginal and Torres Strait Islander Australians 2006* (2007)

Australian Bureau of Statistics, *2076.0 – Census of Population and Housing: Characteristics of Aboriginal and Torres Strait Islander Australians, 2011 – Population Distribution and Structure* (16 February 2018) <www.abs.gov.au/ausstats/abs@.nsf/Lookup/2076.0main+features1102011>

Australian Conservation Foundation, *Submission in Response to the Discussion Paper, National Water Initiative*, April 2004

Australian Government Response to the Report of the Independent Review of the Environment Protection and Biodiversity Conservation Act 1999 (Department of Sustainability, Environment, Water, Population and Communities, 2011)

Brumby, John, 'New Framework a Just Approach to Native Title' (Media Release, 28 July 2010)

Commonwealth, *Parliamentary Debates*, House of Representatives, 16 November 1993

Commonwealth, *Parliamentary Debates*, House of Representatives, 8 August 2007

Commonwealth, *Parliamentary Debates*, Senate, 21 December 1993

Commonwealth, *Parliamentary Debates*, Senate, 16 August 2007

Co-operative Management Agreement between Barengi Gadjin Land Council and State of Victoria (13 December 2005)

Corangamite Catchment Management Authority, *Corangamite Waterway Strategy 2014–2022* (2018)

Council of Australian Governments, *Communiqué* (7 December 1992) <http://archive.coag.gov.au/coag_meeting_outcomes/1992-12-07/index.cfm#environment>

Council of Australian Governments, *Communiqué* (25 February 1994) Attachment A: Water Resource Policy <http://archive.coag.gov.au/coag_meeting_outcomes/1994-02-25/docs/attachment_a.cfm>

Council of Australian Governments, *Communiqué* (25 February 1994) <http://archive.coag.gov.au/coag_meeting_outcomes/1994-02-25/index.cfm>

Council of Australian Governments, *Communiqué* (11 April 1995) Attachment A: Agreement to Implement the National Competition Policy and Related Reforms <http://archive.coag.gov.au/coag_meeting_outcomes/1995-04-11/docs/attachment_a.cfm>

Council of Australian Governments, *Communiqué* (29 August 2003)

Deed of Trust – Ngā Tāngata Tiaki o Whanganui (4 August 2014) <www.ngatāngatatiaki.co.nz/wp-content/uploads/2015/04/Ng%C4%81-T%C4%81ngata-Tiaki-o-Whanganui-Trust-Deed-Final-Executed-Deed-4-August-2014-1.pdf>

Department of Environment and Primary Industries, *Improving Our Waterways — Victorian Waterway Management Strategy* (September 2013)

Department of Environment and Primary Industries, *Meerreeng Wanga: Aboriginal Inclusion Plan 2014–2019* (2014)

Department of Environment, Land, Water and Planning, *Water for Victoria – Water Plan* (2016)

Department of Justice and Department of Sustainability and Environment, *Yorta Yorta Co-operative Management Agreement* (Fact Sheet, May 2004)

Department of Prime Minister and Cabinet (NZ), *Sustainable Development for New Zealand, Program of Action* (January 2003)

Department of the Environment, Water, Heritage and the Arts, *Matters of National Environmental Significance, Significant Impact Guidelines 1.1, Environment Protection and Biodiversity Conservation Act 1999* (2009)

Department of Sustainability and Environment, *Draft Western Region Sustainable Water Strategy for Community Comment* (March 2010)

Department of Sustainability and Environment, *Draft Gippsland Region Sustainable Water Strategy for Community Comment* (August 2010)

Department of Sustainability and Environment, *Indigenous Partnership Framework 2007–10* (2007)

Department of Sustainability and Environment, *A Governance Guide to the Victorian Water Industry* (Victorian Government, East Melbourne, 2011).

Department of Sustainability and Environment, *Gippsland Region Sustainable Water Strategy* (November 2011)

Department of Sustainability and Environment, *Ministerial Guidelines for Assessment of Environmental Effects under the Environment Effects Act 1978* (7th edn, June 2006)

Department of Sustainability and Environment, *Northern Region Sustainable Water Strategy* (November 2009)

Department of Sustainability and Environment, *Sustainable Water Strategy Central Region – Action to 2055* (October 2006)

Department of Sustainability and Environment, *Western Region Sustainable Water Strategy* (November 2011)

East Gippsland Catchment Management Authority, *East Gippsland Waterway Strategy 2014–2022* (2013)

Editorial, *The Argus* (Melbourne), 16 December 1886

Editorial, *The Age* (Melbourne), 17 December 1886

Editorial, *The Age* (Melbourne), 13 January 1881

Editorial, *The Age* (Melbourne), 11 January 1888

Electoral Commission (NZ), *Voter Participation Strategy* (15 July 2013)

Environment Victoria, *Submission in Response to the Discussion Paper, National Water Initiative*, April 2004

Explanatory Memorandum, Native Title Amendment Bill 1997 (Cth)

Federation of Victorian Traditional Owner Corporations, *Victorian Traditional Owner Water Policy Statement 2014* (November 2014)

Federation of Victorian Traditional Owner Corporations, Submission No 72 to the Office of Living Victoria, *Water Law Review*, 14 February 2014

Finlayson, Christopher, 'Whanganui River Agreement Signed' (Press Release, 30 August 2012)

Finlayson, Christopher, 'Waikato River Deed of Settlement Signed with Waikato-Tainui' (Media Release, 17 December 2009)

Finlayson, Christopher, 'Whanganui River Deed of Settlement Initialled' (Media Release, 26 March 2014)

Finlayson, Christopher, 'Whanganui River Deed of Settlement Signed' (Media Release, 5 August 2014)

First Peoples' Water Engagement Council, *Policy Framework* (March 2012)

Frydenberg, Josh, 'Historic Step Towards Global Recognition for Budj Bim' (Media Release, 2 March 2018)

Garrett, Peter, 'EPBC Act Under Independent Review' (Media Release, 31 October 2008)

Glenelg Hopkins Catchment Management Authority, *Glenelg Hopkins Waterway Strategy 2014–2022* (2014)

Glowka, Lyle et al, *A Guide to the Convention on Biological Diversity* (IUCN Environmental Law Centre, 1994)

Godden, Lee, 'Perception of Water in Australian Law: Re-examining Rights and Responsibilities' (Proceedings of the 2003Invitation Symposium: Water – The Australian Dilemma, Academy of Technological Sciences and Engineering, Canberra, 2003)

Goulburn Broken Catchment Management Authority, *Goulburn Broken Regional Waterway Strategy 2013–2021* (2014)

Government Statist, *Census of Victoria, 1891: Parts I to IX, with a General Report by the Government Statist, Taken on the 8th April, 1891* (1893)

'Griffith Protests Over the Murray Darling Basin Plan', *ABC News (Australia)*, 25 January 2011 <www.youtube.com/watch?v=dsXLXHQGq-4>

Her Majesty the Queen in right of New Zealand and Waikato-Tainui, *Deed of Settlement in Relation to the Waikato River* (22 August 2008)

Hoekstra, Gordon, 'Nadleh Whut'en and Stellat'en hereditary leaders proclaim B.C.'s first aboriginal water laws' *Vancouver Sun* (online) 30 March 2016 <http://vancouversun.com/business/energy/nadleh-whuten-and-stellaten-hereditary-leaders-proclaim-b-c-s-first-aboriginal-water-laws>

Howard, John, 'Amended Wik 10 Point Plan' (Press Release, 8 May 1997) <http://pmtranscripts.pmc.gov.au/release/transcript-10333>

Howard, John, 'Transcript of the Prime Minister the Hon. John Howard MP Press Conference – Parliament House' (7 May 1997) <https://pmtranscripts.pmc.gov.au/release/transcript-10331>

Howard, John, *A National Plan for Water Security: 25 January 2007* (Commonwealth of Australia, 2007) <http://pandora.nla.gov.au/pan/10052/20070321-0000/www.pm.gov.au/docs/national_plan_water_security.pdf>

Hulls, Rob, 'Keynote Address' (Speech delivered at the AIATSIS Native Title Conference, Melbourne, 4 June 2009)

Indicator: IW-43 Implementation of National Water Initiative, Department of Sustainability, Environment, Water, Population and Communities (Cth) <http://webarchive.nla.gov.au/gov/20110328054205/http://environment.gov.au/soe/2006/publications/drs/indicator/212/index.html>

Indigenous Advisory Committee, Indigenous Advisory Committee Meeting 25 Bulletin, 23–25 October 2012 Cairns, Queensland (2013), Department of the Environment (Cth), Indigenous Advisory Committee, <www.environment.gov.au/system/files/pages/8ca21d83-fee7-4093-9459-3a6e79e2df15/files/iac-meeting25-bulletin.pdf>

Intergovernmental Agreement on a National Water Initiative (25 June 2004)

Intergovernmental Agreement on the Environment (1 May 1992)

International Union for the Conservation of Nature, *Incorporation of the Rights of Nature as the Organisational Focal Point in IUCN's Decision Making*, Res 100, World Conservation Congress (6–15 September 2012)

Jennings, Gavin, on behalf of the State of Victoria, Gunaikurnai Land & Waters Aboriginal Corporation and Secretary to the Department of Sustainability and Environment, *Traditional Owner Land Management Agreement* (22 October 2010)

King, Tanya, 'Solving Water Security: Don't Ignore the Public' *The Conversation* (online) 3 September 2012 <https://theconversation.com/solving-water-security-dont-ignore-the-public-9173>

Knowles, Craig, Chair, Murray-Darling Basin Authority, Letter to Tony Bourke, Minister for Sustainability, Environment, Water, Population and Communities, 21 November 2012

Macklin, Jenny, 'Statement on the United Nations Declaration of Indigenous Peoples, Parliament House, Canberra, 3 April 2009' <www.un.org/esa/socdev/unpfii/documents/Australia_official_statement_endorsement_UNDRIP.pdf>

Macquarie Dictionary Online (Macquarie Dictionary Publishers, 2018) <www.macquariedictionary.com.au>

Mallee Catchment Management Authority, *Mallee Waterway Strategy 2014–22* (2014)

Melbourne Water, *Diamond Creek – Water Supply Protection Area – Stream Flow Management Plan 2003* (May 2003)

Melbourne Water, *Healthy Waterways Strategy – A Melbourne Water Strategy for Managing Rivers, Estuaries and Wetlands* (November 2013)

Millar, Royce, 'Treasurer Faces Fight Over Pipeline', *The Age* (online), 2 November 2007 <www.theage.com.au/news/national/treasurer-faces-fight-over-pipeline/2007/11/01/1193619061327.html>

Ministry for the Environment, *Your Guide to the Resource Management Act* (August 2006)

Ministry for the Environment, *National Policy Statement for Freshwater Management Amendment Consultation 2013* (13 February 2014) <www.consultation.mfe.govt.nz/content/national-policy-statement-freshwater-management-amendment-consultation-2013>

Ministry for the Environment, *Freshwater Reform 2013 and Beyond* (March 2013)

'Murray Agreement Signed', *Advertiser* (Adelaide), 1 September 1982

Nadleh Whut'en and Stellat'en, *Yinka Dene 'Uza'hné Guide to Surface Water Quality Standards* (18 March 2016) <www.carriersekani.ca/news/yinke-dene-uzahne-guide-to-surface-water-quality-standards>

Nadleh Whut'en and Stellat'en, *Yinka Dene 'Uza'hné Surface Water Management Policy* (18 March 2016) <www.carriersekani.ca/news/yinke-dene-uzahne-guide-to-surface-water-quality-standards>

National Forest Policy Statement – A New Focus for Australia's Forests (1992)

National Native Title Tribunal, *Wotjobaluk, Jaadwa, Jadawadjali, Wergaia and Jupagalk Native Title Determinations – What They Mean for the Wimmera Region* (Information Brochure, December 2006)

National Native Title Tribunal, *The Gundtijmara People's Native Title Determination* (30 March 2007)

National Policy Statement for Freshwater Management 2014 (Updated August 2017)

National Strategy for Ecologically Sustainable Development (December 1992)

National Water Commission, *Position Statement: Indigenous Access to Water Resources* (June 2012)

National Water Commission, *The National Water Initiative – Securing Australia's Water Future: 2011 Assessment* (September 2011)

Nature Conservation Council of New South Wales, *Submission with Regard to National Water Initiative Discussion Paper* (April 2004)

Neville, Lisa, 'New Commissioner to Recognise Aboriginal Water Values' (Media Release, 16 November 2017)

New Zealand, *Parliamentary Debates*, 2 August 2005

New Zealand, *Parliamentary Debates*, 20 April 2010

North Australian Indigenous Experts Water Futures Forum, *Mary River Statement* (6 August 2009)

Northern Australia Indigenous Land and Sea Management Alliance, *A Policy Statement on North Australian Indigenous Water Rights* (November 2009)

Northern Basin Aboriginal Nations, Submission No 72 to the Expert Panel, *Independent Review of the Water Act 2007*, 30 July 2014

North East Catchment Management Authority, *North East Waterway Strategy 2014* (September 2014)

North Central Catchment Management Authority, *2014–22: North Central Waterway Strategy* (2014)

O'Bryan, Katie, Submission No 32 to the Expert Panel, *Independent Review of the Water Act 2007* (4 July 2014)

O'Donnell, Erin, 'Legal Rights for Rivers: More Power Less Protection?' on Gabriel Eckstein, International Water Law Project Blog (23 April 2018) <www.internationalwaterlaw.org/blog>

Office of the Premier (Vic), 'Green Paper Submissions Now Available on the Web' (Media Release, 16 January 2004)

Oxford English Dictionary Online (Oxford University Press, 2018) <www.oed.com>

Port Phillip and Westernport Regional Catchment Strategy, <www.ppwrcs.vic.gov.au./about/indigenous-culture-and-engagement>

Report of the Independent Review of the Water Act 2007 Australian Government Response (nd)

República del Ecuador Constitucion de 2008 [Republic of Ecuador Constitution of 2008] Political Database of the Americas <http://pdba.georgetown.edu/Constitutions/Ecuador/english08.html>

Rotorua Te Arawa Lakes Programme, *Vision and Strategy for the Lakes of the Rotorua District* (2013) <www.rotorualakes.co.nz/strategy>

Settlement Between the Dja Dja Wurrung Traditional Owner Group and the State of Victoria (Fact Sheet, 28 March 2013)

Statistics New Zealand, *2013 Census Quick Stats about Māori* (2013) <http://archive.stats.govt.nz/Census/2013-census/profile-and-summary-reports/quickstats-about-maori-english.aspx>

Streensa, Alex, 'Accommodating Indigenous Cultural Values in Water Resource Management: The Waikato River, New Zealand; the Murray-Darling Basin, Australia; and the Colorado River, USA' (Paper presented at the Australian Agricultural & Resource Economics Society's Annual Conference, Cairns, 11–13 February 2009)

Sullivan, Ann, 'Tōrangapū – Māori and political parties – National, New Zealand First, Māori and Mana parties', *Te Ara: The Encyclopedia of New Zealand* (13 July 2012) <www.TeAra.govt.nz/en/torangapu-maori-and-political-parties/page-4>

Sustainable Water Strategy Central Region – Action to 2055 (DSE, October 2006)

Taupō District Council and Tūwharetoa Māori Trust Board, *Joint Management Agreement* (January 2008)

Te Arawa and Arawa Māori Trust Board and Her Majesty the Queen in right of New Zealand, *Deed of Settlement of the Te Arawa Lakes Historical Treaty Claims and Remaining Annuity Issues* (18 December 2004)

Te Arawa and Arawa Māori Trust Board and Her Majesty the Queen in right of New Zealand, *Schedules to the Deed of Settlement of the Te Arawa Lakes Historical Claims and Remaining Annuity Issues* (18 December 2004)

Te Puni Kōkiri, *He Tirohanga ō Kawa ke te Tiriti o Waitangi. A Guide to the Principles of the Treaty of Waitangi as expressed by the Courts and the Waitangi Tribunal* (Wellington, 2001) <www.tpk.govt.nz/en/a-matou-mohiotanga/crownmaori-relations>

Te Rūnanga o Ngāi Tahu and Department of Conservation, *Te Waihora Joint Management Plan – Mahere Tukutahi o Te Waihora* (10 December 2005)

Thomas, William, Guardian of Aborigines, Letter to Brough Smyth, R, Secretary of the Central Board appointed to watch over the interests of the Aborigines of the Colony of Victoria, 26 July 1860

US Department of State, *Announcement of US Support for the United Nations Declaration on the Rights of Indigenous Peoples* (12 January 2011); <https://2009-2017.state.gov/s/srgia/154553.htm>

Universal Declaration on the Rights of Mother Earth, World People's Conference on Climate Change and the Rights of Mother Earth (22 April 2010), Global Alliance for the Rights of Nature <http://therightsofnature.org/universal-declaration>

Upper Ovens River Water Supply Protection Area: Water Management Plan (11 January 2012)

'Vic Title Claims to Settle out of Court', *The Age* (online), 4 June 2009 <http://news.theage.com.au/breaking-news-national/vic-title-claims-to-settle-out-of-court-20090604-bwnp.html>

Victoria, *Parliamentary Debates, Session 1884 – Legislative Assembly and Legislative Council – Vol XLV*

Victoria, *Parliamentary Debates*, Legislative Assembly, 15 December 1886

Victoria, *Parliamentary Debates*, Legislative Assembly, 23 October 1968

Victoria, *Parliamentary Debates*, Legislative Assembly, 28 October 1970

Victoria, *Parliamentary Debates*, Legislative Assembly, 26 May 1989

Victoria, *Parliamentary Debates*, Legislative Assembly, 6 October 2005

Victoria, *Parliamentary Debates*, Legislative Assembly, 6 April 2006

Victoria, *Parliamentary Debates*, Legislative Assembly, 10 August 2006

Victoria, *Parliamentary Debates*, Legislative Assembly, 15 October 2009

Victoria, *Parliamentary Debates*, Legislative Assembly, 28 July 2010

Victoria, *Parliamentary Debates*, Legislative Assembly, 26 May 2010

Victoria, *Parliamentary Debates*, Legislative Assembly, 31 August 2016

Victoria, *Parliamentary Debates*, Legislative Council, 2 December 1983

Victorian Government, *Victorian Government White Paper – Securing Our Water Future Together* (Department of Sustainability and Environment, June 2004)

Victorian Government, *Water for Victoria: Water Plan* (Department of Environment, Land, Water and Planning, 2016)

Victorian Traditional Owner Land Justice Group, *Statement* (17 February 2005) <www.landjustice.com.au/document/Communique-Statewide-Meeting-17-18Feb05.pdf>

Victorian Traditional Owner Land Justice Group, *Towards a Framework Agreement between the State of Victoria and the Victorian Traditional Owner Land Justice Group* (Discussion Paper, 26 August 2006) <www.landjustice.com.au/document/Framework-Discussion-Paper-26-August-2006.pdf>

Victorian Traditional Owner Land Justice Group, 'Land Justice Within Reach' (Media Release, 28 July 2010)

Waikato Raupatu River Trust and Waikato Regional Council, *Joint Management Agreement* (18 June 2013)

Waikato Raupatu River Trust and Waikato Regional Council, *Co-management Agreement for Waikato River Related Lands* (10 December 2012)

Waikato River Authority, *Waikato River Clean-up Trust Funding Strategy 2015*

Waikato-Tainui and Her Majesty the Queen in right of New Zealand, *Agreement in Principle for the Settlement of the Historical Claims of Waikato-Tainui in relation to the Waikato River* (16 December 2007)

Westbank First Nations Self-Government Agreement 2003

West Gippsland Catchment Management Authority, *West Gippsland Waterway Strategy 2014–2022* (2014)

Whanganui Iwi and the Crown, *Record of Understanding* (13 October 2011) <www.wrmtb.co.nz/new_updates/Record%20of%20Understanding%202012.pdf>

Whanganui Iwi and the Crown, *Ruruku Whakatupua: Te Mana o Te Awa Tupua* (5 August 2014) <www.wrmtb.co.nz/new_updates/Ruruku_Whakatupua_Te_Mana_o_Te_Awa_Tupua_Signed5August%202014.pdf>

Whanganui Iwi and the Crown, *Ruruku Whakatupua: Te Mana o Te Iwi o Whanganui* (5 August 2014) <www.wrmtb.co.nz/new_updates/RurukuWhakatupua-TeManaoTeIwioWhanganui.pdf>

Whanganui Iwi and the Crown, *T⊠tohu Whakatupua* (30 August 2012) <www.wrmtb.co.nz/new_updates/TuutohuWhakatupuaFinalSigned.pdf>

Whanganui River Māori Trust Board, *Whanganui River Settlement: Ratification Results* <www.wrmtb.co.nz/Ratification%20Results%20Panui.pdf>

Wimmera Catchment Management Authority, *Wimmera Waterway Strategy 2014–2022* (2014)

Selected websites

Aboriginal Victoria <www.vic.gov.au/aboriginalvictoria.html>
Council of Australian Governments <www.coag.gov.au>
Department of the Environment and Energy (Cth) <www.environment.gov.au>
Electoral Commission (NZ) <www.elections.org.nz>
Ministry for the Environment (NZ) <www.mfe.govt.nz>
Waitangi Tribunal <www.waitangitribunal.govt.nz>

Index

www.ingramcontent.com/pod-product-compliance
Ingram Content Group UK Ltd.
Pitfield, Milton Keynes, MK11 3LW, UK
UKHW010957180926
475701UK00004B/411

* 9 7 8 0 3 6 7 6 6 4 8 5 5 *